Wireless Networks

Series Editor

Xuemin Sherman Shen, University of Waterloo, Waterloo, ON, Canada

The purpose of Springer's Wireless Networks book series is to establish the state of the art and set the course for future research and development in wireless communication networks. The scope of this series includes not only all aspects of wireless networks (including cellular networks, WiFi, sensor networks, and vehicular networks), but related areas such as cloud computing and big data. The series serves as a central source of references for wireless networks research and development. It aims to publish thorough and cohesive overviews on specific topics in wireless networks, as well as works that are larger in scope than survey articles and that contain more detailed background information. The series also provides coverage of advanced and timely topics worthy of monographs, contributed volumes, textbooks and handbooks.

Haixia Zhang • Dongyang Li • Tong Xue

Intelligent Resources Management for Vehicular Social Networks

Societal Perspectives and Current Issues in the Digital Era

Springer

Haixia Zhang
Shandong University
Jinan, China

Dongyang Li
China University of Petroleum (East China)
Qingdao, China

Tong Xue
Shandong University of Science and Technology
Qingdao, China

ISSN 2366-1186 ISSN 2366-1445 (electronic)
Wireless Networks
ISBN 978-3-031-80168-6 ISBN 978-3-031-80169-3 (eBook)
https://doi.org/10.1007/978-3-031-80169-3

This Springer imprint is published by the registered company Springer Nature Switzerland AG
The registered company address is: Gewerbestrasse 11, 6330 Cham, Switzerland

Preface

One key goal of an intelligent transportation system (ITS) is to ensure the reliability and security of communication among vehicles. False alarms and invalid information transmission can not only misdirect users but also potentially lead to severe traffic accidents. In this area, the integration of vehicle networks and social networks into vehicular social networks (VSNs) is deemed a promising method for improving road safety by leveraging social information to ensure the credibility of data sources. VSNs involve many vehicles and devices as well as massive amounts of communication, computing, and caching resources to accommodate emerging applications and services. These resources are typically limited and can sometimes replace each other to achieve the same system performance. Therefore, resource management is crucial for VSNs. To our knowledge, no book has adequately presented resource management for VSNs. This book is intended to fill this gap and provide academia and industry with comprehensive coverage of resource management for vehicular networks.

Several essential components, such as social awareness models for vehicle behavior prediction, intelligent management of communication, and computing and caching resources, will be elaborated upon in this book, accompanied by in-depth discussions and insightful analysis. This book starts by providing an overview of the key issues encountered in VSNs, including the dynamic and unpredictable nature of vehicular mobility, the varying communication demands involved, and the need for efficient resource allocation. It emphasizes the critical role of resource management in ensuring the seamless functioning of VSNs and its impact on the user experience, safety, and overall network performance. A significant portion of the book is dedicated to exploring behavior prediction models for vehicles in VSNs. Drawing from data analytics and machine learning methodologies, this book presents various prediction algorithms that use historical data to forecast vehicle behavior accurately. By analyzing the past movements, interactions, and social behaviors of vehicles, these models can anticipate future positions, communication patterns, and resource requirements. With behavior prediction as a foundation, the book delves into the design and implementation of intelligent resource management systems for VSNs. It demonstrates how predictive capabilities empower these systems to allocate communication, computing and caching resources dynamically. By anticipating communication needs and potential congestion points, the intelligent resource management system optimizes resource allocation to increase network efficiency, reduce latency, and improve overall network stability. The book extensively evaluates the proposed intelligent resource management approach through extensive simulations and practical experiments. The results demonstrate the effectiveness of the system, highlighting significant improvements in network performance compared with that obtained by traditional resource allocation methods. These findings validate the potential of behavior prediction and intelligent resource management in revolutionizing VSNs. Finally, this book closes with conclusions and promising research directions, aiming to inspire future research in the field of vehicular channels from different perspectives. Although the book is designed primarily for professionals and researchers on wireless communities, advanced students majoring in electrical engineering or computer science may also find the content useful.

Jinan, China Haixia Zhang
Qingdao, China Dongyang Li
Qingdao, China Tong Xue

Contents

Acronyms

AI	Artificial Intelligence
AWFL	Attention-Weighted Federated Learning
AWFL_BiLSTM	Attention-Weighted Federated Learning with a Bidirectional Long Short-Term Memory Network
AVs	Autonomous Vehicles
CAFLPC	Proactive Edge Caching Based on Community Detection and Attention-Weighted Federated Learning
CNN	Convolutional Neural Network
C-V2X	Cellular Vehicle-to-Everything
DSRC	Dedicated Short-Range Communications
FL	Federated Learning
ITS	Intelligent Transportation System
IoVs	Internet of Vehicles
IVs	Important Vehicles
MEC	Mobile Edge Computing
MIMO	Multiple-Input Multiple-Output
NOMA	Nonorthogonal Multiple Access
OBUs	Onboard Units
OMA	Orthogonal Multiple Access
QoS	Quality of Service
RSUs	Roadside Units
STAGCN	Spatial–Temporal Aggregation Graph Convolutional Network
STDenseNet	Spatiotemporal DenseNet
STeP	Spatiotemporal Partial
TWACNet	Time-Wise Attention-Aided Convolutional Neural Network
VSNs	Vehicular Social Networks
V2I	Vehicle-to-Infrastructure

V2P	Vehicle-to-Pedestrian
V2V	Vehicle-to-Vehicle
V2X	Vehicle-to-Everything
VST-BML	Variational Spatial–Temporal Bayesian Metalearning

Chapter 1
Introduction to Vehicular Social Networks

1.1 Overview of Vehicular Social Networks

Over the past few years, the deep integration of sensor and communication technologies has opened up a new design space for in-vehicle applications [1, 2]. In this context, many promising applications, such as autonomous driving, vehicle platooning, and real-time monitoring, have emerged at intervals [3]. With the rapid growth of these smart applications and the advent of advanced communication and computing capabilities, it is possible for drivers and passengers to communicate and socialize with other commuters on the road. Such communication is not limited to interests; these commuters may also interact with other passengers who are facing similar traffic conditions, environmental factors, and mobility patterns or who belong to the same community [4]. The social interdependencies of these commuters can be exploited for diverse applications. Thus, the integration of advanced intelligent computing and social networking perspectives into vehicular networks has emerged as a new paradigm for VSNs and has high potential in diverse areas, not only future intelligent transportation systems (ITSs) but also entertainment and driver comfort on the road [5]. In this monograph, we present in detail the architecture, characteristics, classifications, applications, and challenges of VSNs; more details are given below.

1.1.1 Architecture of VSNs

A VSN is assumed to be a group of individuals who may have common interests, preferences, or needs in the context of temporal and spatial proximity on the road [6]. Specifically, VSNs not only include traditional vehicle-to-vehicle (V2V) and vehicle-to-infrastructure (V2I) communication patterns but are also affected by human factors impacting vehicular connectivity, such as human mobility,

H. Zhang et al., *Intelligent Resources Management for Vehicular Social Networks*, Wireless Networks, https://doi.org/10.1007/978-3-031-80169-3_1

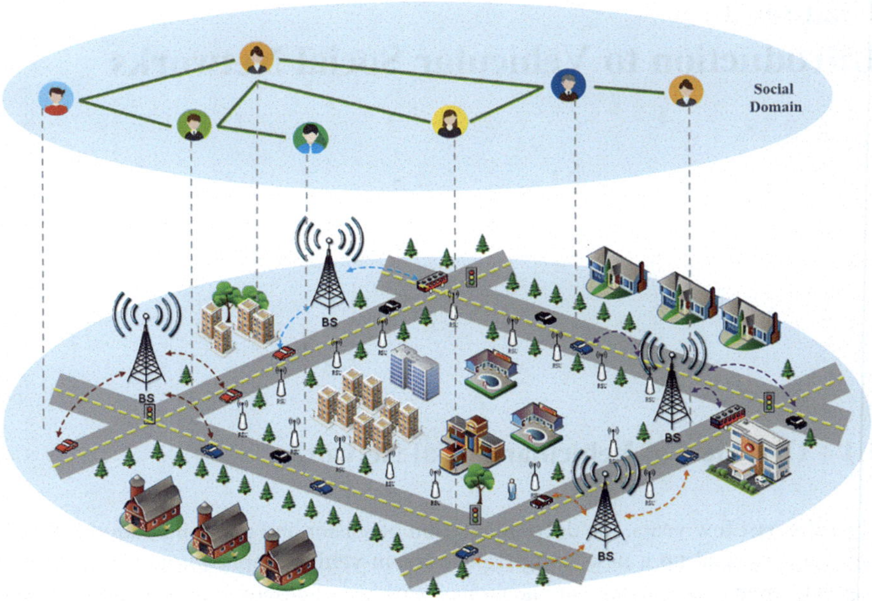

Fig. 1.1 The architecture of VSNs

selfishness, and individual preferences. As shown in Fig. 1.1, the architecture of VSNs is designed to support the complex interactions and communications between V2V and V2I networks and the broader transportation ecosystem. This architecture integrates various components and layers to enable seamless data exchange, efficient traffic management, and enhanced safety and services for passengers and drivers. A detailed overview of the architecture of VSNs is shown below.

1. *Vehicular Nodes*

Vehicular nodes are the fundamental building blocks of VSNs and include all types of vehicles equipped with communication devices and sensors. These nodes interact with each other and with infrastructure nodes to exchange information and coordinate actions. The key aspects of vehicular nodes include the following:

- Onboard Units (OBUs): These units are installed in vehicles and are responsible for collecting, processing, and transmitting data related to vehicle status, location, speed, and operating conditions.
- Sensor Integration: Vehicular nodes may include various sensors, such as GPS receivers, accelerometers, cameras, and environmental sensors (e.g., for air quality monitoring), which are used to gather real-time data about the vehicle's surroundings.
- Communication Interfaces: Vehicular nodes utilize communication technologies such as dedicated short-range communications (DSRC), cellular vehicle-to-

everything (C-V2X), and Wi-Fi (e.g., IEEE 802.11p) to establish wireless links for V2V and V2I communication.

2. *Infrastructure Nodes*

The infrastructure nodes in VSNs consist of fixed installations positioned along roadways and in urban environments. These nodes provide connectivity, data processing capabilities, and support services that enhance the functionality of VSNs. The key infrastructure components include the following:

- Roadside Units (RSUs): When deployed at strategic locations such as intersections, highways, and urban centers, RSUs act as communication gateways between vehicles and the backend infrastructure. They facilitate V2I communication, relay information to vehicles, and support traffic management applications.
- Traffic Management Systems: These systems are integrated with RSUs; they collect and process data from vehicular nodes to monitor traffic flow, analyze congestion patterns, and optimize traffic signal timings on the basis of real-time conditions.
- Cloud and Edge Computing Resources: Backend servers, cloud platforms, and edge computing nodes store, process, and analyze the vast amounts of data generated by VSNs. They support applications such as predictive analytics, traffic simulations, and decision support systems for transportation authorities and service providers.

3. *Communication Protocols and Standards*

Communication protocols are essential for enabling reliable and efficient data exchange within VSNs. Standardized protocols ensure interoperability between different vehicle types, infrastructure providers, and geographic regions. The key communication protocols include the following:

- IEEE 802.11p/WAVE: DSRC standard operating in the 5.9 GHz band, optimized for low-latency V2V and V2I communication.
- C-V2X: Utilizes existing cellular networks (e.g., 4G LTE, 5G) to support V2X communication, providing broader coverage, higher bandwidth, and compatibility with evolving cellular technologies.
- Message Sets and Data Formats: Standardized message sets (e.g., the Basic Safety Message—BSM) define the format and content of data exchanged between vehicular nodes and infrastructure, ensuring consistency and compatibility across different network segments.

4. *Layers of the VSN Architecture*

The architecture of VSNs typically follows a layered approach, similar to traditional network architectures, to manage complexity and ensure efficient operation. The layers include the following:

- Physical Layer: Involves the transmission and reception of wireless signals via communication technologies such as DSRC and C-V2X.
- Data Link Layer: Manages the transmission of data packets between neighboring vehicular nodes, ensuring error-free communication and efficient use of bandwidth.
- Network Layer: Handles routing and forwarding of data packets across the network; it enables dynamic management of network topology and traffic prioritization.
- Transport Layer: Provides end-to-end communication services, ensuring the reliable and secure delivery of data packets between source and destination nodes.
- Application Layer: Implements specific VSN applications and services, such as collision avoidance systems, traffic management algorithms, infotainment services, and environmental monitoring.

5. *Security and Privacy Considerations*

Security and privacy are critical aspects of the VSN architecture, given the sensitivity of the data exchanged and the potential impact of malicious attacks. The key security measures include the following:

- Authentication and Authorization: Verify the identities of the communicating nodes and grant access permissions on the basis of predefined policies.
- Encryption: Secures data transmissions via encryption algorithms to protect against eavesdropping and data tampering.
- Intrusion Detection and Prevention: Monitor network traffic for suspicious activities and implement measures to detect and mitigate cyber threats.
- Privacy-Preserving Techniques: Anonymize sensitive information to protect user identities and personal data while enabling effective communication and data analysis.

6. *Integration with Smart City Initiatives*

VSNs are integral to broader smart city initiatives that aim to improve urban mobility, sustainability, and quality of life. Integration with smart city infrastructure (e.g., IoT sensors, urban data platforms) enables synergy across the transportation, energy, and environmental domains. This integration supports data-driven decision-making, urban planning, and policy formulation in addressing emerging challenges and optimizing resource allocation in urban environments.

The architecture of VSNs is designed to foster seamless communication, enhance traffic management capabilities, improve road safety, and support evolution toward connected and autonomous vehicles. By integrating advanced communication technologies, robust infrastructure, and standardized protocols, VSNs pave the way for innovative transportation solutions that meet the evolving needs of modern cities and communities. Ongoing research and development efforts continue to drive advancements in VSN architecture, addressing challenges and revealing new opportunities for intelligent, efficient, and sustainable transportation systems.

1.1.2 Characteristics of VSNs

VSNs exhibit several distinctive characteristics that define their operational dynamics, functionality, and impact on the transportation ecosystem [7]. These characteristics highlight the way VSNs integrate social interaction with vehicular communication technologies to increase road safety, optimize traffic management, and improve the overall passenger experience. The key characteristics of VSNs are shown in Fig. 1.2.

1. *Integration of Social and Vehicular Elements*

VSNs uniquely combine aspects of social networks with those of vehicular communication systems. They not only facilitate communication among vehicles and between vehicles and infrastructure but also integrate social interactions among drivers, passengers, and pedestrians. This integration supports collaborative applications, community-driven services, and personalized mobility experiences.

2. *Dynamic and Mobile Nodes*

Like traditional vehicular ad hoc networks (VANETs), VSNs involve highly dynamic nodes represented by vehicles moving with varying speeds and directions. The mobility of the nodes leads to dynamic changes in network topology and

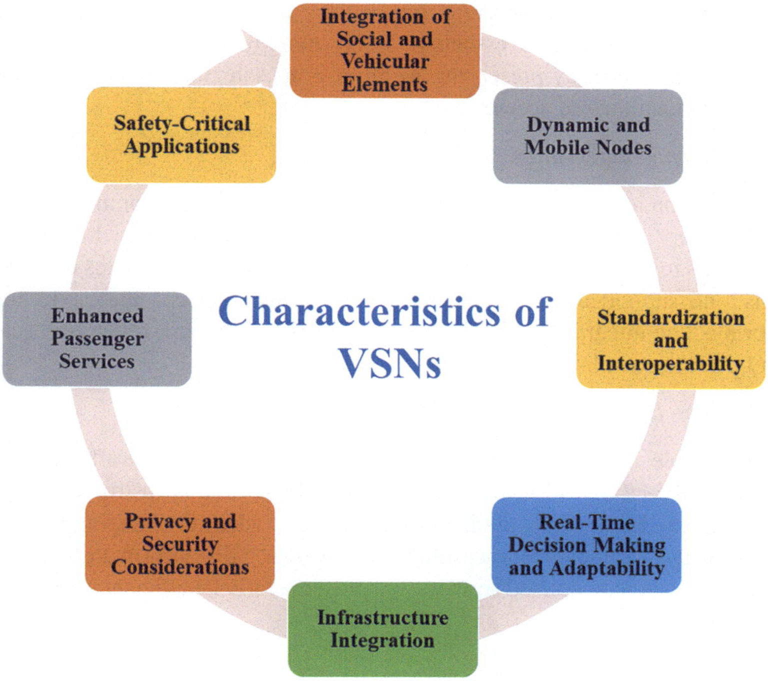

Fig. 1.2 Characteristics of VSNs

connectivity, which influences communication reliability and latency. VSNs require adaptive communication protocols and efficient resource management strategies to maintain connectivity and support diverse applications.

3. *Safety-Critical Applications*

Safety is a primary focus of VSNs and is supported by technology such as collision avoidance systems, cooperative driving assistance, and emergency vehicle notifications. These applications leverage real-time data exchange to increase situational awareness, mitigate risks, and improve overall road safety. VSNs support safety-critical communications with stringent requirements for low latency, reliability, and responsiveness.

4. *Enhanced Passenger Services*

In addition to safety, VSNs offer enhanced passenger services and amenities aimed at improving travel comfort and convenience. These services include real-time traffic updates, navigation assistance, multimedia streaming, and location-based recommendations for nearby services and attractions. VSNs integrate infotainment features to enrich the passenger experience during travel.

5. *Privacy and Security Considerations*

Privacy and security are significant concerns in VSNs because sensitive data, such as location information, driving behaviors, and personal preferences, are collected and exchanged. Protecting user privacy while ensuring secure communication and data integrity is essential. VSNs employ encryption, authentication mechanisms, and privacy-preserving techniques to safeguard data confidentiality and prevent unauthorized access.

6. *Infrastructure Integration*

VSNs rely on a combination of OBUs and RSUs to support communication and service delivery. RSUs are strategically deployed at intersections, highways, and urban centers to extend network coverage, relay data between vehicles and backend systems, and support applications such as traffic management and environmental monitoring. The integration of infrastructure enhances network reliability, scalability, and performance in diverse operational environments.

7. *Real-Time Decision-Making and Adaptability*

VSNs enable real-time decision-making based on dynamic traffic conditions, environmental factors, and social interactions. Vehicles and infrastructure exchange data to optimize traffic flow, predict congestion, and adapt to changing road conditions. Context-aware applications utilize sensor data, GPS information, and social context to deliver personalized services, enhance operational efficiency, and increase user satisfaction.

8. *Standardization and Interoperability*

Standardization is critical to ensuring interoperability and compatibility among VSN components, communication protocols, and service providers. Standardized protocols such as IEEE 802.11p (DSRC) and C-V2X facilitate seamless communication between vehicles, infrastructure, and external systems. Interoperable solutions enable scalable deployment, efficient resource utilization, and broader adoption of VSN technologies across different geographic regions and transportation networks.

The characteristics of VSNs include their unique blend of social interactions, vehicular communication technologies, and advanced networking capabilities. By integrating social elements into transportation systems, VSNs contribute to safer, more efficient, and socially connected mobility environments. Ongoing research and development continue to advance VSN technologies, addressing challenges and expanding opportunities for smart transportation solutions and sustainable urban development initiatives.

1.1.3 Classification of VSNs

VSNs involve communication and interaction between vehicles and infrastructure. These networks serve various purposes, ranging from improving road safety to enabling entertainment and internet access while moving. The following are some classifications of VSNs:

1. *Communication Purpose:*
   - Safety Applications: VSNs often prioritize safety-related communication, such as that related to collision avoidance, road condition monitoring, and emergency alerts.
   - Non-safety Applications: These applications include communication for social purposes, entertainment, internet access, and other noncritical needs.
2. *Network Architecture:*
   - Vehicle-to-Vehicle: Communication occurs directly between vehicles for purposes such as cooperative collision warning and traffic information sharing.
   - Vehicle-to-Infrastructure: Communication occurs between vehicles and fixed infrastructure points such as RSUs or traffic lights, enabling services such as real-time traffic updates and infrastructure-assisted navigation.
   - Vehicle-to-Everything (V2X): This encompasses all types of communication scenarios, including V2V, V2I, and even vehicle-to-pedestrian (V2P) or vehicle-to-network (V2N) communications.
3. *Application Areas:*
   - Safety and Traffic Efficiency: These applications are focused on improving traffic flow, reducing accidents, and enhancing overall road safety.

- Infotainment and Convenience: These services provide entertainment options, internet access, location-based services, and social networking capabilities.
- Commercial Services: The VSN is used for commercial purposes such as fleet management, logistics, and advertising.

4. *Communication Technologies:*
   - Dedicated Short-Range Communication (DSRC): This is a common technology for VSNs; it operates in the 5.9 GHz band and is specifically designed for vehicular communications.
   - Cellular Networks: VSNs are integrated with existing cellular networks (e.g., 4G, 5G) for broader coverage and seamless connectivity.
   - Ad Hoc Networks: Temporary networks are formed among nearby vehicles or between vehicles and infrastructure without relying on a preexisting infrastructure.
5. *Security and Privacy:*
   - Authentication and Access Control: There are measures to ensure that only authorized entities can access the network and its resources.
   - Privacy Preservation: The identities and sensitive information of users and vehicles participating in VSNs are protected.
   - Data Integrity: Transmitted data remain intact and unaltered during transmission.
6. *Geographical Scope:*
   - Urban vs. Rural: VSNs may operate differently depending on whether they are in densely populated urban areas with high vehicle density or in sparsely populated rural areas.
   - Global vs. Local: Some applications may be localized to a specific region, whereas others may require global connectivity (e.g., navigation services).

These classifications highlight the diversity and complexity of VSNs, which continually evolve with advancements in communication technologies and the increasing integration of vehicles into smart city infrastructures [8].

1.1.4 Applications of VSNs

VSNs are formed among vehicles equipped with communication capabilities, typically leveraging technologies such as DSRC and C-V2X. These networks enable vehicles to communicate with each other and with infrastructure, paving the way for several practical applications:

1. *Traffic Management and Routing*

VSNs can provide real-time traffic information to help vehicles avoid congested routes and improve overall traffic flow. This can be achieved through V2V and V2I communication.

2. *Collision Avoidance*

VSNs facilitate the exchange of information about vehicle speed, direction, and proximity, enabling vehicles to warn each other about potential collisions. This can significantly enhance road safety.

3. *Emergency Services*

In the case of accidents or emergencies, VSNs can quickly relay information to emergency responders, improving response times and potentially saving lives.

4. *Infotainment and Passenger Services*

VSNs can enhance the in-vehicle experience by providing entertainment options, internet access, and location-based services to passengers.

5. *Fleet Management*

For businesses managing a fleet of vehicles, VSNs can provide valuable data on vehicle location, status, and performance, enabling logistics and maintenance schedules to be optimized.

6. *Environmental Monitoring*

VSNs can be used to monitor vehicle emissions and air quality in real time, helping municipalities implement effective pollution control measures.

7. *Autonomous Driving*

VSNs play a crucial role in the development of autonomous vehicles by enabling them to communicate with other vehicles and infrastructure to navigate safely and efficiently.

8. *Parking Management*

VSNs can assist drivers in finding available parking spaces, thereby reducing congestion in urban areas and improving the overall efficiency of parking facilities.

9. *Smart City Integration*

VSNs are integral to the concept of smart cities, in which interconnected infrastructure and vehicles work together to optimize transportation systems, improve energy efficiency, and enhance overall urban living standards.

Overall, VSNs are a burgeoning field with vast potential to transform transportation systems, enhance safety, and improve the overall driving and passenger experience through enhanced connectivity and communication capabilities.

1.1.5 Importance of VSNs

VSNs play a pivotal role in modern transportation systems, offering a range of benefits that address key challenges for urban and rural mobility. Several important features highlight their importance:

1. *Enhanced Traffic Management and Efficiency*

VSNs enable real-time communication between vehicles and infrastructure, facilitating efficient traffic management. The applications of such communication include dynamic routing to avoid congestion, optimization of traffic flow through coordinated signaling, and provision of up-to-date information on road conditions. By reducing travel time and fuel consumption, VSNs contribute to economic savings and environmental sustainability.

2. *Improved Safety and Collision Avoidance*

One of the most critical benefits of VSNs is enhancing road safety. Vehicles equipped with VSN technology can exchange information about their speeds, locations, and directions. This enables advanced driver assistance systems (ADASs) and collision avoidance systems to alert drivers to potential hazards and take preventive actions autonomously if necessary. Ultimately, VSNs have the potential to reduce the number of accidents and fatalities on roads.

3. *Facilitation of Autonomous Driving*

VSNs are integral to the development and deployment of autonomous vehicles (AVs). These networks provide essential data for safe AV navigation in complex environments. By sharing information about road conditions, weather updates, and traffic patterns, VSNs improve the decision-making capabilities of AVs, increasing the reliability and feasibility of autonomous driving.

4. *Efficient Parking Management*

Finding parking spaces in crowded urban areas can be a significant challenge. VSNs can address this issue by providing real-time information about available parking spots. Smart parking solutions integrated with VSNs enable drivers to locate nearby parking spaces and reserve them in advance, reducing the traffic congestion caused by drivers searching for parking.

5. *Enhanced Connectivity and Passenger Experience*

VSNs transform vehicles into connected platforms that enhance the overall passenger experience. Passengers can access high-speed internet, stream multimedia content, and interact with social media platforms during their journeys. Furthermore, VSNs enable seamless connectivity between vehicles and smart devices, supporting applications such as vehicle-to-passenger communication and personalized infotainment services.

6. *Environmental Impact and Sustainability*

By optimizing traffic flow and reducing idle time, VSNs contribute to environmental sustainability. Smoother traffic patterns and reduced congestion lead to decreased fuel consumption and emissions. Moreover, VSNs support the integration of electric vehicles by providing real-time data on charging station locations and availability, promoting the adoption of cleaner transportation alternatives.

7. *Support for Smart City Initiatives*

VSNs are a fundamental component of smart city initiatives aimed at improving urban livability and efficiency. By integrating VSN technologies with existing urban infrastructure, cities can improve public transportation systems, implement adaptive traffic management strategies, and promote multimodal transportation solutions. VSNs enable cities to collect data for informed decision-making, monitor air quality levels, and implement sustainable urban planning practices.

8. *Economic Growth and Innovation*

The adoption of VSNs fosters economic growth by creating new opportunities for innovation and entrepreneurship. Companies specializing in VSN technologies, telecommunications, and automotive sectors can develop advanced solutions to solve transportation problems. Furthermore, the deployment of VSNs creates jobs in fields such as technology development, network infrastructure deployment, and data analytics.

In summary, VSNs can revolutionize modern transportation systems by improving traffic management, enhancing road safety, supporting autonomous driving, optimizing parking management, enriching passenger experiences, promoting environmental sustainability, facilitating smart city initiatives, and stimulating economic growth through innovation [9]. As cities and transportation networks continue to evolve, the integration of VSN technologies will play a crucial role in shaping the future of mobility.

1.2 Challenges for Vehicular Social Networks

VSNs face numerous challenges that need to be addressed to enable them to realize their full potential in transforming modern transportation systems. These challenges include resource allocation, vehicle behavior prediction, and other technical challenges. More details are shown in Fig. 1.3.

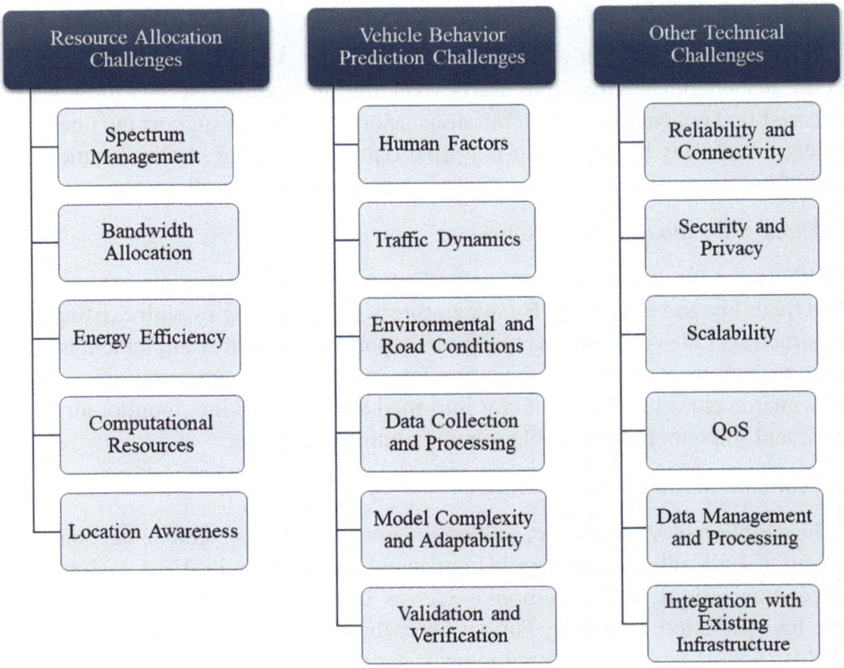

Fig. 1.3 Challenges of VSNs

1.2.1 Resource Allocation Challenges

Resource allocation in VSNs is a critical challenge because of the dynamic and heterogeneous nature of vehicular environments. The key resource allocation challenges and considerations are as follows:

1. *Spectrum Management*

Limited Spectrum Availability: VSNs rely on wireless communication technologies (e.g., DSRC, LTE-V2X) that operate within allocated frequency bands. Spectrum scarcity can limit the scalability and performance of VSNs, especially in densely populated urban areas.

Dynamic Spectrum Access: Vehicles moving at high speeds require efficient spectrum allocation and management methods to maintain connectivity and quality of service (QoS). Dynamic spectrum access techniques, such as spectrum sharing and cognitive radio, are needed to optimize spectrum utilization.

2. *Bandwidth Allocation*

QoS Requirements: Different VSN applications (e.g., real-time traffic updates, video streaming for entertainment) have different QoS requirements (e.g., latency, reliability). Allocating sufficient bandwidth to meet these diverse requirements while ensuring fair access is a challenge.

Traffic Prioritization: Safety-critical applications (e.g., collision avoidance) may require higher priority than non-safety-critical applications. The development of effective traffic prioritization mechanisms is essential to ensure the timely delivery of critical messages.

3. *Energy Efficiency*

Power Consumption: VSNs operate in mobile environments where vehicles rely on battery-powered communication devices. Optimizing energy efficiency is crucial for prolonging battery life and minimizing operational costs.

Dynamic Power Management: Techniques such as sleep modes, adaptive transmission power control, and energy-aware routing are essential for reducing energy consumption without compromising communication performance.

4. *Computational Resources*

Processing and Storage: Vehicles generate large volumes of data that require real-time processing and storage for applications such as traffic management and autonomous driving. Allocating computational resources efficiently while considering vehicle constraints (e.g., processing power, memory) is challenging.

Edge Computing: Leveraging the edge computing capabilities of roadside units or cloudlets can offload computation-intensive tasks from vehicles, reducing latency and increasing scalability in resource-constrained environments.

5. *Location Awareness*

Spatial Resource Allocation: VSNs often require location-aware resource allocation strategies to optimize communication efficiency and minimize interference. Allocating resources on the basis of vehicle density, proximity to infrastructure, and traffic patterns can improve network performance.

Handover Management: Efficient handover mechanisms are necessary when vehicles move between different network segments or communication zones. Seamless handover ensures the continuity of service and minimizes disruptions in resource allocation.

Addressing these resource allocation challenges requires innovative solutions and interdisciplinary approaches involving telecommunications engineering, network management, signal processing, and regulatory frameworks. Advanced algorithms, adaptive strategies, and collaborative efforts among stakeholders are essential to optimize resource allocation and maximize the efficiency and effectiveness of VSNs.

1.2.2 Vehicle Behavior Prediction Challenges

Predicting the behavior of VSNs presents significant challenges due to the complex and dynamic nature of driving environments. These challenges arise from various factors that influence vehicle behavior, including human factors, environmental conditions, and traffic dynamics. The key challenges related to predicting vehicle behavior in VSNs are as follows:

1. *Human Factors*

Driver Intent and Decision Making: Predicting the intentions and decisions of human drivers is challenging because of the variability in individual driving styles, preferences, and cognitive processes.

Driver Distraction and Fatigue: External factors such as distractions (e.g., mobile phones, passengers) and driver fatigue can affect decision-making processes, leading to unpredictable behavior on the road.

Behavioral Adaptation: Drivers may adapt their behavior to real-time traffic conditions, road infrastructure, and social interactions with other drivers, making it challenging to anticipate their actions accurately.

2. *Traffic Dynamics*

Interaction with Surrounding Vehicles: Vehicles interact with each other in complex ways, including overtaking, merging, and yielding behaviors. Predicting these interactions requires understanding the spatial relationships and intentions of multiple vehicles simultaneously.

Traffic Flow Variability: Traffic flow patterns can vary significantly on the basis of factors such as congestion levels, road geometry, and traffic signals. Predicting how these factors influence vehicle behavior requires real-time data and advanced modeling techniques.

3. *Environmental and Road Conditions*

Weather and Visibility: Adverse weather conditions (e.g., rain, snow) and poor visibility can impact vehicle behavior, affecting factors such as speed, following distances, and maneuvering capabilities.

Road Surface Conditions: Changes in road surface conditions (e.g., wet roads, potholes) can influence vehicle handling and stability, necessitating adaptive prediction models that account for environmental factors.

4. *Data Collection and Processing*

Data Availability and Accuracy: Predictive models rely on accurate and real-time data inputs, including vehicle trajectories, speed profiles, and sensor readings. Ensuring data availability from diverse sources (e.g., vehicle sensors, roadside units) and maintaining data accuracy are crucial.

Real-Time Processing: Predicting vehicle behavior in real time requires efficient data processing and analysis, considering the volume of data generated by vehicles and infrastructure in VSNs and the speed of data generation.

5. *Model Complexity and Adaptability*

Modeling Uncertainty: Uncertainties in predicting human behavior and environmental conditions necessitate probabilistic modeling approaches that account for uncertainty in predictions.

Adaptability to Dynamic Environments: VSNs operate in dynamic environments where conditions can change rapidly. Predictive models must be adaptable to evolving traffic situations, unforeseen events (e.g., accidents, road closures), and emerging driver behaviors.

6. *Validation and Verification*

Model Validation: Validating predictive models against real-world data and scenarios is challenging due to the complexity and variability of driving environments. Ensuring the reliability and accuracy of predictions requires rigorous testing and validation processes.

Addressing these challenges requires interdisciplinary approaches that integrate advanced data analytics, machine learning techniques, human factor research, and real-time sensor technologies. The development of robust predictive models for vehicle behavior in VSNs is essential for enhancing traffic management, increasing safety, and optimizing the overall efficiency of transportation systems.

1.2.3 Other Technical Challenges

Technical challenges in VSNs arise from the unique characteristics of vehicular environments: vehicles are mobile, communication conditions are dynamic, and diverse applications require varying levels of reliability and real-time interaction. Some key technical challenges faced by VSNs are as follows:

1. *Reliability and Connectivity*

Interference and Signal Strength: Maintaining reliable communication among vehicles and between vehicles and infrastructure is challenging because of factors such as signal interference due to nearby vehicles, buildings, and environmental conditions (e.g., weather).

Mobility Management: Vehicles moving at high speeds require seamless handovers with base stations or roadside units (RSUs) to maintain continuous connectivity. Managing mobility-related issues, such as frequent network entry and exit, is crucial for uninterrupted communication.

2. *Security and Privacy*

Data Integrity and Authentication: Ensuring the authenticity and integrity of messages exchanged in VSNs is essential to prevent malicious attacks, such as message spoofing or tampering.

Privacy Concerns: VSNs involve the collection and sharing of sensitive data, including location information and driving behaviors. The implementation of robust privacy-preserving techniques to protect user identities and personal information while enabling effective communication is critical.

3. *Scalability*

Handling Large-Scale Deployment: As the number of connected vehicles increases, VSNs must support many simultaneous connections and manage network resources efficiently to avoid congestion and maintain performance.

Network Load Balancing: Efficiently distributing network loads and balancing resource allocation across vehicles and infrastructure nodes (e.g., RSUs) is challenging in dynamic vehicular environments with varying traffic densities.

4. *QoS*

Real-Time Communication Requirements: Many VSN applications, such as collision avoidance systems and real-time traffic updates, require stringent QoS guarantees, including low latency and high reliability. Ensuring consistent QoS across diverse applications and network conditions is complex.

Traffic Prioritization: Differentiating between safety-critical and non-safety-critical traffic to prioritize the transmission and timely delivery of critical messages is crucial for enhancing safety and efficiency in VSNs.

5. *Data Management and Processing*

Big Data Challenges: Large volumes of data are generated in VSNs by vehicles, infrastructure sensors, and connected devices. Efficiently managing, processing, and analyzing these data in real time to extract actionable insights for applications such as traffic management and predictive maintenance is a significant challenge.

Edge Computing and Fog Computing: Leveraging edge computing and fog computing paradigms to perform computations closer to the data source (e.g., a vehicle or an RSU) reduces latency, conserves bandwidth, and enhances the scalability of VSNs.

6. *Integration with Existing Infrastructure*

Legacy System Compatibility: Integrating VSN technologies with existing transportation infrastructure, including legacy communication systems and roadside infrastructure, requires interoperability and backward compatibility to facilitate smooth transition and deployment.

Standardization: The lack of standardized protocols and interfaces across different VSN technologies and geographical regions complicates integration efforts and hinders seamless communication and collaboration between vehicles and infrastructure.

Addressing these technical challenges requires collaborative efforts among researchers, engineers, policymakers, and industry stakeholders to develop innovative solutions, standards, and regulatory frameworks. Overcoming these challenges will enable VSNs to realize their full potential to enhance transportation safety, efficiency, and the overall mobility experience.

1.3 Resource Management in Vehicular Social Networks

Owing to the open nature of VSNs, untrusted vehicles are abundant [10]. When information is transmitted to untrusted or irrelevant vehicles, the reliability of the computed results will significantly decrease. This may lead to a re-offloading or re-transmission process and consume more resources, thus degrading the system performance of VSNs, such as the user experience, system energy efficiency, or spectrum efficiency. On the basis of [11–14], the VSN is regarded as a key technology for reducing the probability of re-offloading and retransmission by associating users with nodes of interest or trusted nodes through the social relationships among users. However, with an increasing number of vehicles, it will be difficult to accommodate the rapidly growing demand for communication services through limited resources. Investigating spectrum resource allocation, computation offloading, and content caching to improve overall resource utilization is essential for addressing these challenges in VSNs, and these ideas have attracted considerable attention from academia and industry [15–19]. In this section, we examine existing works related to resource management in VSNs, including communication resource management schemes, computational resource allocation strategies, caching resource management schemes, and multidimensional resource allocation strategies.

1.3.1 Communication Resource Management

To address the issues caused by the increasing demand for vehicular data transmission and the limited amount of communication resources (such as spectrum resources and power resources), many existing works have investigated the communication resource management problem for VSNs. To efficiently manage the overall communication resources of VSNs and achieve high resource utilization, several advanced communication resource reuse techniques have been introduced. In what follows, the existing communication resource management schemes for VSNs are summarized from four perspectives, i.e., orthogonal multiple access (OMA)-based communication resource management, nonorthogonal multiple access (NOMA)-based communication resource management, multiple-input multiple-output (MIMO)-based communication resource management, and MIMO-NOMA-based communication resource management.

1. *OMA-Based Communication Resource Management*

In VSNs, the spectrum resource reuse mode between a V2I link and a V2V link can be classified as underlaid and overlaid. For overlaid-based VSNs, V2V users occupy dedicated spectrum resources that should have been assigned to V2I users [20, 21]. With dramatic increases in the number of vehicles and variety of vehicular applications, severe channel competition between V2I users and V2V users has arisen. It is increasingly difficult for overlaid-based VSNs to serve a large number of vehicles with limited bandwidth resources, and this has restricted network system performance. To address this issue, the underlaid-based VSN has been extensively studied; in such VSNs, V2V users share bandwidth resources with V2I users [22, 23]. In this way, many users can be accommodated. However, this also causes severe cochannel interference between V2I users and V2V users. If this problem is not properly solved, the network system performance will be greatly decreased, and the QoS requirements of V2V users and V2I users may not be satisfied.

2. *NOMA-Based Communication Resource Management*

It is essential to study efficient spectrum access strategies for underlaid-based VSNs to ensure that the QoS requirements of each vehicle user are met. Many works have focused on this topic [24–28], and a series of joint spectrum and power decision algorithms that optimize system performance have been studied. Recently, NOMA technology has been adopted to satisfy the ultrahigh sum rate and QoS requirements of vehicular users [29]. By utilizing power domain multiplexing, NOMA enables multiple users to reuse the subchannel to transmit signals simultaneously [30–32]. With successive interference cancellation (SIC) technology, cochannel interference can be suppressed, as shown in Fig. 1.4, ultimately leading to a dramatic improvement in the system sum-rate performance [33–35]. Many NOMA-enabled resource management strategies have been investigated in non-safety service transmission scenarios [36], safety service transmission scenarios [37, 38] and multiple-service transmission scenarios [39, 40]. To optimize the sum transmission rate of safety services, [41] designed a machine learning-based bandwidth and power allocation algorithm to roughly meet both the high-reliability requirements of safety

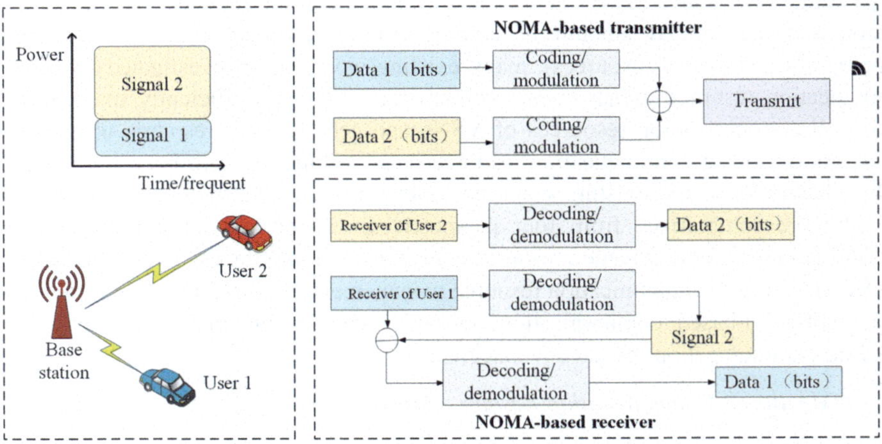

Fig. 1.4 NOMA scheme applying SIC in vehicular networks

services and the system transmission rate requirements of underlaid-based VSNs. In [42], with the same objective, a NOMA-based graph-enabled spectrum assignment approach and a noncooperative game power control algorithm were designed to support multiple types of services. [43] designed a robust spectrum assignment and power allocation strategy to maximize the system sum rate when there is an imperfect channel state.

3. *MIMO-Based Communication Resource Management*

MIMO is another technology that provides high spectrum efficiency in 5G/B5G or 6G networks [44]. A large-scale antenna array can transmit massive data streams with different spatial patterns concurrently at the same frequency. Hence, throughput within the same spectrum is greatly enhanced. There are several other challenges in obtaining an optimal MIMO transmission design under the aforementioned considerations for a VSN: (i) MIMO beamforming design for a vehicular network is a typical nonconvex optimization problem, and (ii) many real-life factors, such as feedback channel noises, time latencies, and frequency offsets in wireless channels, will result in imperfect channel-state information. More importantly, some characteristics of vehicle scenarios play a significant and unique role. For example, the impact of vehicular mobility and the pattern of traffic flows on roads are recognized as the main factors influencing the characteristics of a VSN. Vehicular mobility can lead to a dynamically changing spectrum availability as perceived by vehicles, which, in turn, makes it more challenging to obtain accurate sensing information on cognitive-to-primary propagation channels. Vehicular motion may also cause time variation in the propagation channels, impacting the correctness of the estimation of secondary vehicle-to-vehicle channels. In addition, the absence or impracticality of centralized computation increases the complexity of MIMO transmission design in vehicular scenarios. Therefore, an appropriate MIMO beamforming design solution needs to capture inaccuracies in the information of vehicular channels to be sufficiently robust for online distributed implementation so that it can accommodate the dynamic nature of propagation channels in vehicular scenarios to some extent. To date, many studies have focused on the energy efficiency (EE) of transmission in MIMO systems, such as those in [45–49]. The authors of [45] aimed to achieve a high EE and proposed an EE strategy for V2I and infrastructure-to-infrastructure communications on the basis of cooperative relay and cooperative MIMO transmissions. Although their method was proven to be more effective than the traditional single-input–single-output transmission strategy, it did not consider some real-life requirements, such as the maximum power budget and the required data rate. In [46], the researchers considered obtaining an optimal tradeoff between the sum of data rates and the power cost for MIMO broadcast channels.

4. *MIMO-NOMA-Based Communication Resource Management*

The combination of massive MIMO and NOMA technologies provides a promising solution for achieving extremely high spectral efficiency in 5G. However, massive MIMO-NOMA systems present several major technical challenges. First, the MIMO RF chain elements, including the RF amplifier and analog-to-digital/

digital-to-analog converter, increase the system complexity and hardware cost. One critical technique for operating MIMO-NOMA with limited RF chains is to select the best subset of antennas for these RF chains [50–52]. Furthermore, since channel conditions vary over time and frequency, it is more cost-effective to schedule users into appropriate channels at each time slot. Moreover, NOMA requires user pairing, which typically involves one nearby user (with strong channel gain) and one far user (with weak channel gain) being scheduled as a user pair for SIC. Hence, it is important to solve the antenna selection and user scheduling problem jointly in massive MIMO-NOMA systems. Few researchers have investigated this aspect of MIMO-NOMA systems. In [53], user pairing algorithms were proposed for two NOMA systems, but only single-band scenarios were considered. A proportional fairness scheduler was employed to determine user pairs in [54]. B. Kim et al. scheduled two users with high correlation and large channel gain differences as a user pair [55]. However, these two methods focus only on the NOMA user pairing scheme. A norm-based joint selection scheme for transmit and receive antennas was presented in [56], which exhaustively searches all candidate transmit and receive antennas to find a subset with the largest channel norm. In [57], a joint antenna selection and user scheduling algorithm was proposed. It successively deletes the undesired antennas and users that make a minimum contribution to system performance. However, these two algorithms can be applied only to scenarios with small sets of candidate antennas and users because of their high computational complexity. It is difficult to apply these algorithms to massive MIMO-NOMA systems. To solve this problem, Liu et al. [58] investigated the antenna selection and user scheduling problems in a massive MIMO-NOMA system. They first considered a simple single-band two-user scenario, for which an efficient search algorithm for antenna selection was proposed. A limited number of candidate antennas are examined to find the desired antennas, which are beneficial to the related users. Then, they proposed a joint antenna–user contribution algorithm for joint antenna selection and user scheduling in a multiband, multiuser scenario. It jointly considers the contribution of each antenna's and user's channel gain to the total channel gain.

1.3.2 Computation Resource Management

This section surveys current research related to decisions regarding computation offloading to a server. The papers are divided into those that consider only full offloading and those that consider the possibility of partial offloading.

1. *Full Offloading*

These works focus on full offloading to minimize the execution delay, minimize the energy consumption of the vehicle while satisfying a predefined delay constraint, or find a proper trade-off between the energy consumption and execution delay.

- *Minimization of the Execution Delay:*

The goal of minimizing the execution delay was pursued in [59]. A one-dimensional search algorithm was proposed, which determines an optimal offloading decision policy according to the application buffer queuing state, available processing power of the vehicle and of the mobile edge computing (MEC) server, and characteristics of the channel between the vehicle and the MEC server. The computation offloading decision itself is made on the vehicle side by means of a computation offloading policy module. This module decides, during each time slot, whether an application waiting in the buffer should be processed locally or on the MEC server while minimizing the execution delay. Another idea aimed at minimizing the execution delay was introduced in [60]. Like the previous study, [61] also reduced application failure for offloaded applications. The paper assumed that the user would apply dynamic voltage and frequency scaling [62] and energy harvesting techniques [63] to minimize energy consumption during local execution and would use power control to optimize data transmission for computation offloading. In this context, the authors proposed a low-complexity Lyapunov optimization-based dynamic computation offloading algorithm. It makes offloading decisions in each time slot and subsequently allocates CPU cycles for the vehicle or allocates transmission power.

- *Minimization of Energy Consumption While Satisfying an Execution Delay Constraint:*

A computation offloading method that minimized the energy consumption at the user equipment (UE) while satisfying the execution delay requirement of the application was proposed in [64]. An extension of [65] from a single-UE scenario to a multi-UE scenario was considered in [66]. Another offloading decision strategy for the multi-UE case that aimed to minimize the energy consumption of the users while satisfying the maximum allowed execution delay requirement was proposed in [67]. The main weak point of [67] is that it assumes only a single node, and consequently, interference among the UEs connected to various nodes is not considered. Hence, the work in [67] was extended in [68] to a multicell scenario with N nodes to reflect real network deployment. Since the optimization problem formulated in [67] was no longer convex, the authors proposed a distributed iterative algorithm exploiting successive convex approximation that converged to a local optimal solution.

- *Trade-Off Between Energy Consumption and Execution Delay:*

The computation offloading decision for a multiuser multichannel environment considering the trade-off between the energy consumption at the UE and the execution delay was proposed in [69]. The extension of [69] from a single-UE scenario to a multi-UE scenario was presented in [70]. Since the multiple UEs were assumed to be connected to the same computing node, the offloading decision was made jointly with the allocation of computing and communication resources to all UEs. Analogous to [70], the proposal in [71] outperformed the strategy in which computation is always performed by the UE and the strategy in which computation is always offloaded to the server.

2. *Partial Offloading*

This subsection focuses on works dealing with partial offloading. We classify these as studies focused on minimizing the energy consumption of the vehicle while satisfying the predefined delay constraint and studies that find a proper trade-off between the energy consumption and the execution delay.

- *Minimization of Energy Consumption While Satisfying the Execution Delay Constraint:*

[66] considered applications divided into a non-offloading part and *N* offloading parts. The main objective of the paper was to decide which offloading parts should be offloaded to the MEC server. Minimizing energy consumption while satisfying the delay constraints of the whole application was also the main objective of [66]. In contrast to [67], the application scenario in [68] was assumed to involve several atomic parts that are dependent on each other. [69] addressed the partial offloading decision problem for the multi-UE scenario. In [70], the application to be offloaded does not contain any non-offloading parts, and in some extreme cases, the whole application may be offloaded if this is profitable. In [71], the energy consumption was minimized while satisfying the execution delay of the application. To do this, the authors proposed an energy-optimal partial offloading scheme that forces the user to adapt its computing power depending on the maximal allowed latency of the application.

- *Trade-Off Between Energy Consumption and Execution Delay:*

A trade-off analysis between the energy consumption and the execution delay for making partial offloading decisions was performed in [72]. The application to be offloaded contains only offloading parts, and in extreme cases, full offloading may occur. The offloading decision involves the following parameters: the total number of bits to be processed, the computational capabilities of the user and MEC, the channel state between the user and the server that provides access to MEC, and the energy consumption of the user. The computation offloading decision is formulated as the joint optimization of communication and computational resource allocation. The study in [73] provides a more in-depth theoretical analysis of the trade-off between the energy consumption and the latency of the offloaded applications, which was preliminarily handled in [73]. The main drawback of [74] is that it considered only the single-user scenario. A trade-off analysis between the energy consumption on the user side and the execution delay for the multiuser scenario was presented in [75]. In a multi-UE scenario, the joint optimization process proposed in [76] must be further modified since both the communication and computation resources provided by the MEC server are shared among multiple users. The trade-off between the power consumption and the execution delay for the multiuser scenario was also addressed.

If a decision regarding the full or partial offloading of an application to the MEC server is made (as discussed in the previous section), proper allocation of the computational resources must be performed. Similarly, to the computation

Table 1.1 Comparison of individual papers addressing the allocation of computational resources

	Computing nodes	Objective	Proposed solution
[77]	MEC Cloud server	Maximize number of served applications Satisfy delay constraint	Priority-based cooperation policy
[78]	MEC Cloud server	Minimize delay Minimize power	Optimal online application assignment policy using equivalent discrete MDP framework
[79]	Cloud server	Minimize delay	Optimal allocation policy obtained by solving MDP using linear programing reformulation
[80]	Vehicle MEC Cloud server	Balance communication and computation loads of computing nodes Minimize resource utilization	Online approximation algorithms with polynomial–logarithmic competitive ratio for tree application graph placement

offloading decision, the selection of computation placement is influenced by the ability of the offloaded application to be parallelized/partitioned. If parallelization/partitioning of the application is not possible, only one physical node may be allocated for computing since the application cannot be split into several parts. In the opposite case, the offloaded application may be processed by resources distributed over several computing nodes. This section reviews the papers addressing the problem of proper allocation of computing resources for applications that will be offloaded to the MEC server. We categorize the related works in Table 1.1.

1.3.3 Caching Resource Management

Various studies have proposed different system architectures for content delivery in vehicular networks. Dzyiauddin et al. [81] surveyed computational offloading, content delivery, and caching in vehicular edge computing, including the architecture, communication layers, and applications of vehicular edge computing for content delivery and caching. Zhang et al. [82] introduced a socially aware mobile edge computing architecture for content caching. They employed the DRL model and proposed a method of exploiting the relationships among vehicles and RSUs to perform content dissemination with diverse vehicular social characteristics for urban environments. Moreover, they extended their model by introducing digital twin technology to map the edge caching system into cyberspace [83]. They used the vehicular cloud to coordinate the correlation of cached content among multiple vehicles and then employed a deep learning approach for route selection, accounting for the social context, vehicular cloud formation and cache resource allocation. Similarly, Qiao et al. [84] introduced a cooperative vehicular edge caching system to optimize content delivery and placement within a vehicular edge computing environment with the aid of flexible cooperation between cellular stations, RSUs, and vehicular nodes. The optimization problem was formulated as a double-time-scale Markov decision process (DTS-MDP). Zhou et al. [85] introduced a new content delivery architecture by utilizing 5G edge networks, and content caching and data prefetching methods were discussed. Furthermore, they studied the comprehensive dynamic link utilization problem for 5G edge networks from the perspectives of network operators and vehicle users. Luo et al. [86] introduced an intelligent algorithm (EdgeVCD) based on a content distribution mechanism. It uses a dual-importance (DI) evaluation method to reflect the relationship between the priority of vehicles (PoV) and the priority of content (PoC) and formulates an optimization problem to maximize the system utility for content distribution. De Souza et al. proposed a safe and sound (SNS) approach, which uses a hybrid architecture and cooperative rerouting method to increase system performance and scalability [87]. SNS uses a recurrent neural network (RRN) to predict future safety risk dynamics and to perform customized rerouting in which every vehicle chooses risks to avoid. When the traffic server detects a congested road, it notifies the incoming vehicles by sending a traffic report to all vehicles whose paths cross this road segment. The objective of the SNS rerouting strategy is to balance the traffic flow over a set of alternative routes for each vehicle on the basis of its current and final positions and preferences. Soua et al. proposed a VSN architecture that combines the content-centric networking (CCN) model and software-defined networking (SDN) to offer a multipronged approach for adaptive content delivery [88]. Each content item includes the location and name of the requester. Intermediate nodes that receive the requester message check it in their local content store (CS). If the requested content is not available in the CS, they forward the message and trigger a timer. The FC is used to support geographic content routing. An SDN controller operates on a direct path between the content requester and the provider, similar to the dynamic unicast method. Alowish et al. proposed a content delivery architecture for vehicular networks called Cuckoo, in which the RSU delivers user content via controller nodes [89]. In other words, the RSU selects the optimal routing path to the provider location with the help of the controller nodes. Moreover, Zhao et al. proposed DP-IB, a vehicular content delivery system that uses data pouring and a buffering mechanism for content dissemination in VANETs. The source node sends data content buffered along the selected path and rebroadcasts it at road intersections. The main idea behind data pouring (DP) is that instead of broadcasting the content to the entire network, the system only sends the disseminated content to a few road segments, known as axis roads (A-roads). Usually, A-roads have denser traffic flow than other roads and are chosen as main roads. Intersection buffering (IB) means that content is disseminated to nodes traveling along the crossing roads (C-roads) intersecting the A-roads. To address the privacy leakage problem in social vehicular networks, Zhang et al. introduced a distributed location privacy-preserving spatial crowdsourcing method for the IoV [90]. It enables vehicular nodes to be involved in spatial crowdsourcing and guarantees the privacy of location information. They employed blockchain to record user data without requiring a centralized spatial crowdsourcing server.

1.3.4 Multidimensional Resource Management

Considering the coupled characteristics of communication, computing, and storage resources, it is important to design multidimensional resource management schemes for jointly managing these resources, as this has the potential to significantly improve overall resource utilization in VSNs [91–93]. Zhou et al. [94] designed a caching, computing, and communication scheme for edge caching-based video streaming to maximize the network's EE. However, [95] focused mainly on energy consumption unilaterally. It was difficult for the proposed method to achieve optimal system performance, especially from the perspective of power allocation. In [96], Li et al. aimed to minimize the weighted sum of the execution delay and energy consumption in MEC systems, and partial offloading scheduling and power allocation schemes were jointly considered. However, the caching and computing capabilities of edge nodes were not considered in [97]. The advantages of MEC in reducing latency and energy consumption could not be fully utilized. Motivated by the analysis above, Liu et al. [98] investigated a delay and energy optimization problem in a VSN scenario, where caching, computing, and power allocation were jointly considered.

Although cooperation among vehicles can effectively facilitate vehicular content sharing, selecting cooperative partners for each vehicle remains a problem. VSNs have been extensively studied to help vehicles safely and efficiently exchange infotainment information with others by leveraging the social relationships among vehicles [99–101]. To classify vehicles through social analysis, Lin et al. considered historical and current social correlations on the basis of the number of identical road segments [102]. Moreover, community similarity criteria, such as similar movements and privacy ratings of vehicles, were considered to design a dynamic probability caching scheme in [103]. Vegni et al. reported that mobility leads to differences between VSNs and online social networks (OSNs) [104]. [104] noted that a VSN connects users on the basis of the encounter metric, as opposed to a traditional social network, which links users via offline friendships. However, none of these studies on social similarity considered the randomness of vehicle encounters. Therefore, traditional social relationships based on communities [105] can hardly be applied to vehicular communication networks. It is assumed that different vehicles meet randomly, and if two specific vehicles meet more often, the possibility of cooperation between them is greater, which means that the social relationship between them is stronger. In addition to social relationships, content popularity is an important factor affecting content sharing efficiency. In this work, we propose a new content caching strategy based on vehicles' social similarity and content popularity to facilitate content caching and content delivery.

Despite the advantages of VSNs, user selfishness is still a factor because of the additional resource consumption in VSNs. In [106], the authors reported that for non-security-related applications in VSNs, the selfish behavior of nodes can degrade network performance. Additionally, selfish users want to pay less for a higher QoS and may keep their QoS requirements private. As a result, the problem of asymmetric information arises, which will lead to inefficient resource allocation since the content provider will be uncertain about the content requesters' QoS requirements and may uniformly schedule resources with an undifferentiated price. To address this challenge, contract theory can be applied as a formal framework to study the interactions between rational players under asymmetric information, avoiding the effect of users' fraudulent behaviors [107]. [108] focused on incentive mechanism design to address the content sharing problem of BS control and proposed the inclusion of a performance reward in the contract to encourage content providers to invest more effort. Xu et al. [109] investigated the content sharing problem for urban vehicular networks. To increase content sharing efficiency and address information asymmetry issues, communication, computing, and storage resources are jointly allocated, and vehicle mobility is considered. However, joint communication, computing, and storage resource allocation for all content requesters and content providers in the considered scenario is complex and intractable since the allocation processes of different resources have different time scales. Note that content caching is a long-term process (i.e., has a large time scale) since users' content preferences usually evolve relatively slowly, i.e., on the order of hours or even days.

1.4 Organization of the Monograph

In this monograph, we investigate intelligent resource management issues in VSNs. In Chap. 2, we propose a novel framework for wireless traffic prediction in VSNs, where the movement behavior and communication behavior of vehicles are considered jointly. On the basis of the prediction results, Chap. 3 presents a social-mobility-aware radio communication resource allocation algorithm to improve the sum-rate performance in VSNs. In Chap.4, a socially aware proactive edge caching strategy for VSNs is presented to achieve a trade-off between the cost of providing caching services and the content transmission latency. In Chap. 5, we design a joint communication and computational resource management strategy to improve the offloading EE and propose a trusted relay-based scheme. The details of this monograph are summarized as follows.

In Chap. 2, we first propose a novel method to match the movement and communication behavior of vehicles by merging two independent datasets concerning the trajectories of vehicles and communication traffic volumes. Then, a novel STeP-UNet is proposed, in which STeP is embedded to capture cross-domain features of the wireless traffic pattern, and the UNet structure is utilized to determine the skip connections from the front layer to the back layer to fuse different resolutions.

In Chap. 3, a social-mobility-aware underlying C-V2X network is investigated. To meet the various QoS requirements of both cellular users and vehicular users and maximize the system sum rate, we propose jointly optimizing vehicle pairing and radio resource (spectrum and power) assignment. Since the formulated optimization problem is NP-hard, an iterative method is proposed to solve it by decoupling the original problem into two subproblems, i.e., the joint vehicle pairing and spectrum assignment subproblem and the power control subproblem. The HSMA-VPA and a revised KM-SAA are designed to solve the first subproblem. Given the results of vehicle pairing and spectrum assignment, the second subproblem is solved via low-complexity 3D PCA. Through iterations, the vehicle pairing variables, spectrum assignment variables, and transmit power control are iteratively updated until convergence.

In Chap. 4, a novel community detection and attention-weighted federated learning-based proactive edge caching strategy is proposed. In this strategy, we first group vehicles into different communities on the basis of both the mobility and social properties of the vehicles and then select IVs as caching nodes for each community by considering the social importance of the vehicles. To determine how to place popular content in these selected IVs, an AWFL-based content popularity prediction framework is proposed. It integrates AWFL and BiLSTM to achieve higher content popularity prediction accuracy while protecting user privacy. Considering the imbalance of the vehicles' active levels and local computing capacities, an attention-weighted aggregation mechanism is proposed to improve the training efficiency and prediction accuracy.

In Chap. 5, we focus on social-mobility-aware EE maximization in underlaid C-V2X networks, where the T-VUs can offload computational tasks to the S-VUs and the RSUs. An EE maximization problem is formulated to jointly assign the 2C resources. Since the optimization process is an MINLP problem, we transform it into separate computation and communication resource allocation subproblems. To address the first subproblem, we fully integrate social and mobility characteristics and design a heuristic algorithm to perform edge server selection and task splitting. To address the complex cochannel interference in the second subproblem, the power allocation and spectrum assignment solutions are obtained via a tightening lower bound method and a KM algorithm. The simulation results show that the proposed JCCRAA can not only prevent task offloading to untrusted users but also improve resource utilization and system EE.

In Chap. 6, we investigate the secrecy rate performance of wireless-powered FD relay systems. By adopting the TS protocol, the FD relay harvests energy from the source RF signals and stores it as the power supply for confidential information transmission. The objective of this method is to maximize the system secrecy rate. In doing so, an optimization problem is formulated by jointly considering the energy beamformer, the information beamformer, and the TS coefficient. As the formulated problem is nonconvex, to solve it, an iterative algorithm is proposed. The closed-form solutions of the energy beamformer and information beamformer are derived theoretically. Moreover, on the basis of the concavity of the secrecy rate with respect to the TS coefficient, a numerical solution is obtained. Simulations were also performed to verify the proposed scheme, and the results reveal that the choice of the TS coefficient plays a critical role in determining the wireless-powered FD relay gains. In addition, it is shown that exploiting two antennas of the FD relay for energy harvesting can increase the amount of harvested energy, thereby enhancing the secrecy rate performance.

Finally, we conclude this monograph and discuss potential future research directions in Chap. 7, including context-aware resource allocation, autonomous resource management, edge computing and edge intelligence, artificial intelligence (AI) for resource optimization, federated learning for resource optimization, privacy-preserving resource management, interoperability and standardization, simulation, and testbed development.

1.5 Summary

In this chapter, we provided a comprehensive overview of current VSNs, including their architecture, characteristics, classification, and applications. Then, the major challenges faced by VSNs were investigated, followed by a brief survey on resource management in VSNs. Finally, the organization of this monograph was presented.

References

1. Boualouache, A., & Engel, T. (2023). A survey on machine learning-based misbehavior detection systems for 5G and beyond vehicular networks. *IEEE Communications Surveys & Tutorials, 25*(2), 1128–1172., Second quarter.
2. Chen, X., et al. (2024). Vehicle as a service (VaaS): Leverage vehicles to build service networks and capabilities for smart cities. *IEEE Communications Surveys & Tutorials.* https://doi.org/10.1109/COMST.2024.3370169
3. Kuutti, S., Fallah, S., Katsaros, K., Dianati, M., Mccullough, F., & Mouzakitis, A. (2018). A survey of the state-of-the-art localization techniques and their potentials for autonomous vehicle applications. *IEEE Internet of Things Journal, 5*(2), 829–846.
4. Rahim, A., Kong, X., Xia, F., et al. (2018). Vehicular social networks: A survey. *Pervasive and Mobile Computing, 43*, 96–113.
5. Hua, L. C., Anisi, M. H., et al. (2017). Social networking-based cooperation mechanisms in vehicular ad-hoc network-a survey. *Vehicular Communications, 10*, 57–73.
6. Lu, R. (2012). *Security and privacy preservation in vehicular social networks*. Ph.D. dissertation, Electrical and Computer Engineering, University Waterloo.
7. Aung, N., et al. (2023). VeSoNet: Traffic-aware content caching for vehicular social networks using deep reinforcement learning. *IEEE Transactions on Intelligent Transportation Systems, 24*(8), 8638–8649.
8. Fan, K., et al. (2020). A secure and verifiable data sharing scheme based on blockchain in vehicular social networks. *IEEE Transactions on Vehicular Technology, 69*(6), 5826–5835.
9. Ning, Z., Xia, F., Ullah, N., Kong, X., & Hu, X. (2017). Vehicular social networks: Enabling smart mobility. *IEEE Communications Magazine, 55*(5), 16–55.
10. Luo, X., Wen, J., Kang, J., Nie, J., Xiong, Z., Zhang, Y., Yang, Z., & Xie, S. (2023). Privacy attacks and defenses for digital twin migrations in vehicular metaverses. *IEEE Network*, 1–1.
11. Zhang, Y., et al. (2024). An efficient caching and offloading resource allocation strategy in vehicular social networks. *IEEE Transactions on Vehicular Technology, 73*(4), 5690–5703.

12. Zhang, L., Zhao, Z., Wu, Q., Zhao, H., Xu, H., & Wu, X. (2018). Energy-aware dynamic resource allocation in UAV assisted Mobile edge computing over social internet of vehicles. *IEEE Access, 6*, 56700–56715.
13. Hu, B., & Chu, X. (2021). Social-aware resource allocation for vehicle-to-everything communications underlaying cellular networks. In *2021 IEEE 93rd vehicular technology conference (VTC2021-Spring), Helsinki, Finland* (pp. 1–6).
14. Zhang, Y., & Zhou, Y. (2023). Resource allocation strategy based on tripartite graph in vehicular social networks. *IEEE Transactions on Network Science and Engineering, 10*(5), 3017–3031.
15. Adam, M. M., Zhao, L., Wang, K., & Han, Z. (2023). Beyond 5G networks: Integration of communication, computing, caching, and control. *China Communications, 20*(7), 137–174.
16. Zhou, Y., Yu, F. R., Chen, J., & Kuo, Y. (2018). Communications, caching, and computing for next generation HetNets. *IEEE Wireless Communications, 25*(4), 104–111.
17. Ndikumana, A., et al. (2020). Joint communication, computation, caching, and control in big data multi-access edge computing. *IEEE Transactions on Mobile Computing, 19*(6), 1359–1374.
18. Karabulut Kurt, G., & Yanikomeroglu, H. (2021). Communication, computing, caching, and sensing for next-generation aerial delivery networks: Using a high-altitude Platform Station as an enabling technology. *IEEE Vehicular Technology Magazine, 16*(3), 108–117.
19. Tang, M., Gao, L., & Huang, J. (2020). Communication, computation, and caching resource sharing for the internet of things. *IEEE Communications Magazine, 58*(4), 75–80.
20. Hu, B., & Chu, X. (2021). Social-aware resource allocation for vehicle-to-everything communications underlaying cellular networks. In *2021 IEEE 93rd vehicular technology conference (VTC2021-Spring)* (pp. 1–6).
21. Tian, J., Liu, Q., Zhang, H., & Wu, D. (2022). Multiagent deep-reinforcement learning-based resource allocation for heterogeneous qos guarantees for vehicular networks. *IEEE Internet of Things Journal, 9*(3), 1683–1695.
22. Fan, B., He, Z., Wu, Y., He, J., Chen, Y., & Jiang, L. (2020). Deep learning empowered traffic offloading in intelligent software defined cellular v2x networks. *IEEE Transactions on Vehicular Technology, 69*(11), 13328–13340.
23. Aslani, R., Saberinia, E., & Rasti, M. (2020). Resource allocation for cellular v2x networks mode-3 with underlay approach in lte-v standard. *IEEE Transactions on Vehicular Technology, 69*(8), 8601–8612.
24. Liang, L., Ye, H., Yu, G., & Li, G. Y. (2020). Deep-learning-based wireless resource allocation with application to vehicular networks. *Proceedings of the IEEE, 108*(2), 341–356.
25. Guo, C., Liang, L., & Li, G. Y. (2019). Resource allocation for low-latency vehicular communications: An effective capacity perspective. *IEEE Journal on Selected Areas in Communications, 37*(4), 905–917.
26. Hu, X., Xu, S., Wang, L., Wang, Y., Liu, Z., Xu, L., Li, Y., & Wang, W. (2021). A joint power and bandwidth allocation method based on deep reinforcement learning for v2v communications in 5g. *China Communications, 18*(7), 25–35.
27. Wen, S., Liang, W., Cui, J., Wang, D., & Li, L. (2021). Resource allocation based on three-sided matching theory in cognitive vehicular networks. In *2021 IEEE 94th vehicular technology conference (VTC2021-Fall)* (pp. 1–6).
28. Xue, T., Zhang, H., Ding, H., & Yuan, D. (2021). Vehicle pairing, spectrum assignment and power control for non-safety critical services in heterogeneous vehicular networks. *IEEE Transactions on Vehicular Technology, 70*(9), 9160–9173.
29. Situ, Z., Ho, I. W.-H., Hou, Y., & Li, P. (2020). The feasibility of noma in c-v2x. In *IEEE INFOCOM 2020 – IEEE conference on computer communications workshops (INFOCOM WKSHPS)* (pp. 562–567).
30. Dong, P., Ning, Z., Ma, R., Wang, X., Hu, X., & Hu, B. (2020). Noma-based energy-efficient task scheduling in vehicular edge computing networks: A self-imitation learning-based approach. *China Communications, 17*(11), 1–11.

31. Sun, Y., Ding, Z., Dai, X., Navaie, K., & So, D. K. C. (2020). Performance of downlink noma in vehicular communication networks: An analysis based on poisson line cox point process. *IEEE Transactions on Vehicular Technology, 69*(11), 14001–14006.
32. Pei, X., Yu, H., Chen, Y., Wen, M., & Chen, G. (2020). Hybrid multicast/unicast design in Noma-based vehicular caching system. *IEEE Transactions on Vehicular Technology, 69*(12), 16304–16308.
33. Chen, L., Ma, L., Xu, Y., & Leung, V. C. M. (2019). Hypergraph spectral clustering based spectrum resource allocation for dense noma-hetnet. *IEEE Wireless Communications Letters, 8*(1), 305–308.
34. Wen, X., Zhang, H., Zhang, H., & Fang, F. (2019). Interference pricing resource allocation and user-subchannel matching for Noma hierarchy fog networks. *IEEE Journal of Selected Topics in Signal Processing, 13*(3), 467–479.
35. Li, R., Hong, P., Xue, K., Zhang, M., & Yang, T. (2021). Resource allocation for uplink noma-based d2d communication in energy harvesting scenario: A two-stage game approach. *IEEE Transactions on Wireless Communications*, 1–15.
36. Wu, W., Xue, T., Wang, Q., Han, S., & Wu, X. (2019). A two-stage resource allocation for scma-based c-v2x networks. In *2019 IEEE global communications conference (GLOBECOM)* (pp. 1–6).
37. Ding, H., & Leung, K.-C. (2021). Resource allocation for low-latency noma v2x networks using reinforcement learning. In *IEEE INFOCOM 2021 – IEEE conference on computer communications workshops (INFOCOMWKSHPS)* (pp. 1–6).
38. Di, B., Song, L., Li, Y., & Li, G. Y. (2017). Non-orthogonal multiple access for high-reliable and low-latency v2x communications in 5g systems. *IEEE Journal on Selected Areas in Communications, 35*(10), 2383–2397.
39. Zhang, F., Wang, M. M., Bao, X., & Liu, W. (2021). Centralized resource allocation and distributed power control for noma-integrated nr v2x. *IEEE Internet of Things Journal, 8*(22), 16522–16534.
40. Guo, S., & Zhou, X. (2019). Robust resource allocation with imperfect channel estimation in noma-based heterogeneous vehicular networks. *IEEE Transactions on Communications, 67*(3), 2321–2332.
41. Zhang, F., Wang, M. M., Chen, S., Shan, L., Xiao, M., & Xu, M. (2021). Applying noma to nr v2x: A graph-based matching and cooperative game approach. In *2021 IEEE 93rd vehicular technology conference (VTC2021-Spring)* (pp. 1–5).
42. Kim, J., Joung, J., & Lee, J. W. (2018). Resource allocation for multiple device-to-device cluster multicast communications underlay cellular networks. *IEEE Communications Letters, 22*(2), 412–415.
43. Yu, K., Zhou, H., Qian, B., Tang, Z., & Shen, X. (2020). A reinforcement learning aided decoupled ran slicing framework for cellular v2x. In *GLOBECOM 2020–2020 IEEE global communications conference* (pp. 1–6).
44. Jungnickel, V., Manolakis, K., Zirwas, W., Panzner, B., Braun, V., Lossow, M., Sternad, M., Apelfröjd, R., & Svensson, T. (2014). The role of small cells, coordinated multipoint, and massive MIMO in 5G. *IEEE Communications Magazine, 52*(5), 44–51.
45. Nguyen, T.-D., Berder, O., & Sentieys, O. (2011). Energy-efficient cooperative techniques for infrastructure-to-vehicle communications. *IEEE Transactions on Intelligent Transportation Systems, 12*(3), 659–668.
46. Xu, J., & Qiu, L. (2013). Energy efficiency optimization for MIMO broadcast channels. *IEEE Transactions on Wireless Communications, 12*(2), 690–701.
47. Jiang, C., & Cimini, L. (2013). Energy-efficient transmission for MIMO interference channels. *IEEE Transactions on Wireless Communications, 12*(6), 2988–2999.
48. Zhong, W., & Wang, J. (2013). Energy efficient spectrum sharing strategy selection for cognitive MIMO interference channels. *IEEE Transactions on Signal Processing, 61*(14), 3705–3717.

49. Zhang, X., Li, H., Lu, Y., & Zhou, B. (2015). Distributed energy efficiency optimization for MIMO cognitive radio network. *IEEE Communications Letters, 19*(5), 847–850.
50. Gao, X., Edfors, O., Liu, J., & Tufvesson, F. (2013). Antenna selection in measured massive MIMO channels using convex optimization. In *Proceedings of the IEEE GC workshops* (pp. 129–134).
51. Hanif, M., Juntti, M., & Tran, L. (in press). Antenna selection with erroneous covariance matrices under secrecy constraints. *IEEE Transactions on Vehicular Technology.*
52. Amadori, P. V., & Masouros, C. (2015). Power efficient massive MU-MIMO via antenna selection for constructive interference optimization. In *Proceedings of the IEEE ICC* (pp. 1607–1612).
53. Ding, Z., Fan, P., & Poor, H. V. (in press). Impact of user pairing on 5G nonOrthogonal multiple access. *IEEE Transactions on Vehicular Technology.*
54. Benjebbour, A., Li, A., Kishiyama, Y., Jiang, H., & Nakamura, T. (2014). System-level performance of downlink NOMA combined with MIMO for future LTE enhancements. In *2014 IEEE Globecom workshops conference (GC Wkshps)* (pp. 706–710).
55. Kim, B., Lim, S., Kim, H., Suh, S., Kwun, J., Choi, S., Lee, C., Lee, S., & Hong, D. (2014). Non-orthogonal multiple access in a downlink multiuser beamforming system. In *MILCOM 2014 IEEE (MILCOM)* (pp. 706–710).
56. Zhang, P., Chen, S., & Hanzo, L. (2014). Two-tier channel estimation aided near-capacity MIMO transceivers relying on norm-based joint transmit and receive antenna selection. *IEEE Transactions on Wireless Communications, 14*(1), 122–137.
57. Benmimoune, M., Driouch, E., Ajib, W., & Massicotte, D. (2015). Joint transmit antenna selection and user scheduling for Massive MIMO systems. In *2015 IEEE wireless communications and networking conference (IEEE WCNC 2015)* (pp. 381–386).
58. Liu, X., & Wang, X. (2016). Efficient antenna selection and user scheduling in 5G massive MIMO-NOMA system. In *2016 IEEE 83rd vehicular technology conference (VTC Spring), Nanjing, China* (pp. 1–5).
59. Liu, J., Mao, Y., Zhang, J., & Letaief, K. B. (2016). Delay-optimal computation task scheduling for mobile-edge computing systems. In *IEEE international symposium on information theory (ISIT), Barcelona, Spain* (pp. 1451–1455).
60. Mao, Y., Zhang, J., & Letaief, K. B. (2016). Dynamic computation offloading for mobile-edge computing with energy harvesting devices. *IEEE Journal on Selected Areas in Communications, 34*(12), 3590–3605.
61. Zhang, W., et al. (2013). Energy-optimal mobile cloud computing under stochastic wireless channel. *IEEE Transactions on Wireless Communications, 12*(9), 4569–4581.
62. Ulukus, S., et al. (2015). Energy harvesting wireless communications: A review of recent advances. *IEEE Journal on Selected Areas in Communications, 33*(3), 360–381.
63. Kamoun, M., Labidi, W., & Sarkiss, M. (2015). Joint resource allocation and offloading strategies in cloud enabled cellular networks. In *IEEE international conference on communications (ICC), London, U.K.* (pp. 5529–5534).
64. Labidi, W., Sarkiss, M., & Kamoun, M. (2015). Joint multi-user resource scheduling and computation offloading in small cell networks. In *IEEE wireless mobile computing, networking and communications (WiMob), Abu Dhabi, UAE* (pp. 794–801).
65. Barbarossa, S., Sardellitti, S., & Di Lorenzo, P. (2013). Joint allocation of computation and communication resources in multiuser mobile cloud computing. In *IEEE workshop signal processing advances wireless communications (SPAWC), Darmstadt, Germany* (pp. 26–30).
66. Sardellitti, S., Scutari, G., & Barbarossa, S. (2014). Joint optimization of radio and computational resources for multicell mobile cloud computing. In *Proceedings of the IEEE international workshop signal processing advances wireless communications (SPAWC), Toronto, ON, Canada* (pp. 354–358).
67. Chen, X., Jiao, L., Li, W., & Fu, X. (2016). Efficient multi-user computation offloading for mobile-edge cloud computing. *IEEE/ACM Transactions on Networking, 24*(5), 2795–2808.

68. Chen, M.-H., Liang, B., & Dong, M. (*2015*). A semidefinite relaxation approach to mobile cloud offloading with computing access point. In *2015 IEEE international workshop on signal processing advances in wireless communications (SPAWC), Stockholm, Sweden* (pp. 186–190).
69. Chen, M.-H., Dong, M., & Liang, B. (2016). Joint offloading decision and resource allocation for mobile cloud with computing access point. In *International conference on acoustics, speech, and signal processing (ICASSP), Pudong, China* (pp. 3516–3520).
70. Cao, S., Tao, X., Hou, Y., & Cui, Q. (2015). An energy-optimal offloading algorithm of mobile computing based on HetNets. In *International conference on connected vehicles and expo (ICCVE), Shenzhen, China* (pp. 254–258).
71. Deng, M., Tian, H., & Fan, B. (2016). Fine-granularity based application offloading policy in cloud-enhanced small cell networks. In *IEEE international conference on communications workshops (ICC), Kuala Lumpur, Malaysia* (pp. 638–643).
72. You, C., & Huang, K. (2016). Multiuser resource allocation for mobile edge computation offloading. In *IEEE conference on global communications (GLOBECOM), Washington, DC, USA* (pp. 1–6).
73. Wang, Y., Sheng, M., Wang, X., Wang, L., & Li, J. (2016). Mobile-edge computing: Partial computation offloading using dynamic voltage scaling. *IEEE Transactions on Communications, 64*(10), 4268–4282.
74. Muñoz, O., Pascual-Iserte, A., & Vidal, J. (2013). Joint allocation of radio and computational resources in wireless application offloading. In *Future network and mobile summit, Lisbon, Portugal* (pp. 1–10).
75. Muñoz, O., Pascual-Iserte, A., & Vidal, J. (2015). Optimization of radio and computational resources for energy efficiency in latency-constrained application offloading. *IEEE Transactions on Vehicular Technology, 64*(10), 4738–4755.
76. Muñoz, O., Pascual-Iserte, A., Vidal, J., & Molina, M. (2014). Energy-latency trade-off for multiuser wireless computation offloading. In *IEEE conference on wireless communications and networking (WCNCW), Istanbul, Turkey* (pp. 29–33).
77. Zhao, T., Zhou, S., Guo, X., Zhao, Y., & Niu, Z. (2015). A cooperative scheduling scheme of local cloud and Internet cloud for delay-aware mobile cloud computing. In *IEEE Globecom workshops (GC Wkshps), San Diego, CA, USA* (pp. 1–6).
78. Guo, X., Singh, R., Zhao, T., & Niu, Z. (2016). An index based task assignment policy for achieving optimal power-delay tradeoff in edge cloud systems. In *IEEE international conference on communications (ICC), Kuala Lumpur, Malaysia* (pp. 1–7).
79. Di Valerio, V., & Presti, F. L. (2014). Optimal virtual machines allocation in mobile femto-cloud computing: An MDP approach. In *IEEE conference on wireless communications and networking (WCNCW), Istanbul, Turkey* (pp. 7–11).
80. Wang, S., Zafer, M., & Leung, K. K. (2017). Online placement of multi-component applications in edge computing environments. *IEEE Access, 5*, 2514–2533.
81. Dziyauddin, R. A., et al. (2021). Computation offloading and content caching and delivery in vehicular edge network: A survey. *Computer Networks, 197*.
82. Zhang, K., Cao, J., Liu, H., Maharjan, S., & Zhang, Y. (2020). Deep reinforcement learning for social-aware edge computing and caching in urban informatics. *IEEE Transactions on Industrial Informatics, 16*(8), 5467–5477.
83. Zhang, K., Cao, J., Maharjan, S., & Zhang, Y. (2022). Digital twin empowered content caching in social-aware vehicular edge networks. *IEEE Transactions on Computational Social Systems, 9*(1), 239–251.
84. Qiao, G., Leng, S., Maharjan, S., Zhang, Y., & Ansari, N. (2020). Deep reinforcement learning for cooperative content caching in vehicular edge computing and networks. *IEEE Internet of Things Journal, 7*(1), 247–257.
85. Zhou, H., Cheng, N., Wang, J., Chen, J., Yu, Q., & Shen, X. (2019). Toward dynamic link utilization for efficient vehicular edge content distribution. *IEEE Transactions on Vehicular Technology, 68*(9), 8301–8313.

86. Luo, Q., Li, C., Luan, T. H., & Shi, W. (2020). EdgeVCD: Intelligent algorithm inspired content distribution in vehicular edge computing network. *IEEE Internet of Things Journal, 7*(6), 5562–5579.
87. de Souza, A. M., Braun, T., Botega, L. C., Villas, L. A., & Loureiro, A. A. F. (2020). Safe and sound: Driver safety-aware vehicle re-routing based on spatiotemporal information. *IEEE Transactions on Intelligent Transportation Systems, 21*(9), 3973–3989.
88. Soua, R., et al. (2017). SDN coordination for CCN and FC content dissemination in VANETs. In *Ad Hoc networks* (pp. 221–233). Springer.
89. Alowish, M., Shiraishi, Y., Takano, Y., Mohri, M., & Morii, M. (2020). A novel software-defined networking controlled vehicular named-data networking for trustworthy emergency data dissemination and content retrieval assisted by evolved interest packet. *International Journal of Distributed Sensor Networks, 16*(3).
90. Zhao, J., Zhang, Y., & Cao, G. (2007). Data pouring and buffering on the road: A new data dissemination paradigm for vehicular ad hoc networks. *IEEE Transactions on Vehicular Technology, 56*(6), 3266–3277.
91. Markakis, E. K., Karras, K., Sideris, A., Alexiou, G., & Pallis, E. (2017). Computing, caching, and communication at the edge: The cornerstone for building a versatile 5G ecosystem. *IEEE Communications Magazine, 55*(11), 152–157.
92. Zhou, J., Wu, F., Zhang, K., Mao, Y., & Leng, S. (2018). Joint optimization of offloading and resource allocation in vehicular networks with mobile edge computing. In *International conference on wireless communications and signal processing (WCSP), Hangzhou, China* (pp. 1–6).
93. Guo, H., Liu, J., Ren, J., & Zhang, Y. (2020). Intelligent task offloading in vehicular edge computing networks. *IEEE Wireless Communications, 27*(4), 126–132.
94. Zhou, G., Zhao, L., Wang, Y., Zheng, G., & Hanzo, L. (2020). Energy efficiency and delay optimization for edge caching aided video streaming. *IEEE Transactions on Vehicular Technology, 69*(11), 14116–14121.
95. Li, L., Kuang, Z., & Liu, A. (2019). Energy efficient and low delay partial offloading scheduling and power allocation for mec. In *Proceedings of the IEEE International Conference Communications* (pp. 1–6).
96. Liu, W., Zhang, H., Ding, H., & Yuan, D. (2022). Delay and energy minimization for adaptive video streaming: A joint edge caching, computing and power allocation approach. *IEEE Transactions on Vehicular Technology, 71*(9), 9602–9612.
97. Alam, K. M., Saini, M., & Saddik, A. E. (2015). Toward social Internet of vehicles: Concept, architecture, and applications. *IEEE Access, 3*, 343–357.
98. Zhang, B., Tian, R., & Li, C. (2020). Content dissemination and routing for vehicular social networks: A networking perspective. *IEEE Wireless Communications, 27*(2), 118–126.
99. Sun, W., Liu, J., Yue, Y., & Jiang, Y. (2020). Social-aware incentive mechanisms for D2D resource sharing in IIoT. *IEEE Transactions on Industrial Informatics, 16*(8), 5517–5526.
100. Lin, K., Li, C., Fortino, G., & Rodrigues, J. J. (2018). Vehicle route selection based on game evolution in social Internet of vehicles. *IEEE Internet of Things Journal, 5*(4), 2423–2430.
101. Zhao, W., Qin, Y., Gao, D., Foh, C. H., & Chao, H.-C. (2017). An efficient cache strategy in information centric networking vehicle-to-vehicle scenario. *IEEE Access, 5*, 12657–12667.
102. Vegni, A. M., & Loscrí, V. (2015). A survey on vehicular social networks. *IEEE Communications Surveys and Tutorials, 17*(4), 2397–2419., 4th Quart.
103. Vegni, A. M., Loscri, V., & Benslimane, A. (2020). SOLVER: A framework for the integration of online social networks with vehicular social networks. *IEEE Networks, 34*(1), 204–213.
104. Mohaien, A., Kune, D. F., Vasserman, E. Y., Kim, M., & Kim, Y. (2013). Secure encounter-based mobile social networks: Requirements, designs, and tradeoffs. *IEEE Transactions on Dependable and Secure Computing, 10*(6), 380–393.
105. Zhang, B., Li, Y., Jin, D., Hui, P., & Han, Z. (2015). Social-aware peer discovery for D2D communications underlaying cellular networks. *IEEE Transactions on Wireless Communications, 14*(5), 2426–2439.

106. Nitti, M., Girau, R., Floris, A., & Atzori, L. (2014). On adding the social dimension to the Internet of vehicles: Friendship and middleware. In *IEEE international Black Sea conference on communications and networking (BlackSeaCom), Odessa, Ukraine* (pp. 134–138).
107. Bolton, P., & Dewatripont, M. (2005). *Contract theory*. MIT Press.
108. Zhang, T., Wang, H., & He, J. (2016). An incentive mechanism under hidden-action for device-to-device content sharing. In *IEEE industrial electronics society (ICSP), Chengdu, China* (pp. 1288–1292).
109. Xu, L., et al. (2022). Socially driven joint optimization of communication, caching, and computing resources in vehicular networks. *IEEE Transactions on Wireless Communications, 21*(1), 461–476.

Chapter 2
Learning-Based Vehicle Behavior Prediction in VSNs

Wireless traffic prediction plays an important role in developing more intelligent and greener communication networks [1]. By gathering and analyzing historical data, one can predict the future trend of wireless traffic in a network and preassign the necessary resources on demand. The various quality-of-service (QoS) requirements of vehicles can be satisfied with high resource utilization efficiency [2, 3]. Accurate wireless traffic prediction relies on an understanding of the communication behaviors inherent in historical data, where the spatiotemporal distribution of communication demands is expected to be captured. Deep learning [4–7], which was originally used in areas such as computer vision (CV) and natural language processing (NLP), is considered a promising tool for wireless traffic prediction [8–10]. For example, an adaptive traffic prediction problem was formulated from a probabilistic inference perspective in [8], and a variational spatial-temporal Bayesian meta-learning (VST-BML) algorithm was developed to model the nonlinear spatial-temporal dependencies of wireless traffic. The authors in [9] proposed a temporal attention-aided convolutional neural network (TWACNet) structure for cellular traffic prediction. In the proposed TWACNet, the temporal attention mechanism was adopted to capture the long-range temporal dependencies of cellular traffic data, and a convolutional neural network (CNN) was adopted to capture spatial correlations. In [11], an efficient spatiotemporal identity-based multigraph convolutional network framework (SIMF) was proposed for conducting application-level traffic prediction in the metaverse. In the SIMF, a spatiotemporal embedding layer and a multigraph convolutional module were designed to jointly capture the spatiotemporal correlations among nodes and reduce the dependence on topological information, making the situation more consistent with the real relationships between avatars in the metaverse. Moreover, [12] proposed a spatial–temporal aggregation-based graph convolutional network (STAGCN), in which the daily historical patterns and the hourly current–day patterns of mobile traffic are modeled. Moreover, complex spatial–temporal correlations are captured by an aggregation graph convolutional

H. Zhang et al., *Intelligent Resources Management for Vehicular Social Networks*, Wireless Networks, https://doi.org/10.1007/978-3-031-80169-3_2

network for all nodes across different timestamps. The impact of the external factors on mobile traffic is fed into a regression module in the last step to obtain the predicted traffic [13]. investigated the spatial and temporal dependencies of wireless traffic among different cells, where a spatiotemporal DenseNet (STDenseNet)-based prediction framework was designed.

Although promising methods have been reported in the literature for predicting communication traffic in traditional wireless networks, they are rarely applied directly to the Internet of Vehicles (IoV) scenario. In the IoV, the network topology changes dynamically owing to the movements of users (i.e., vehicles). The spatio-temporal distribution of communication traffic is therefore impacted not only by the imposed communication demands but also by the movement behaviors of the vehicles. The correlation between them should be considered, as it is critical to the resulting prediction performance. In fact, several studies have considered vehicle traffic [14, 15]. For example, [14] examined a combination consisting of the ConvPlus and SemanticPlus modules, where the former captured long-range spatial dependencies and the latter was employed to reduce the impacts of the selected location function on crowd movement. In [15], the authors regarded the dynamics of traffic data as a temporal evolution scenario. With the assistance of ensemble learning, three typical graph-level prediction methods were employed to construct an integrated and enhanced forecasting model. This design uses several methods to minimize the uncertainty in the CIoV. However, to our knowledge, few efforts have been made to consider the joint prediction of vehicle traffic and wireless communication traffic in the context of the IoV.

In this chapter, we propose a novel framework for performing wireless traffic prediction in the IoV, where the movement and communication behaviors of vehicles are jointly considered. To achieve this goal, we first propose a pattern matching approach for merging a wireless traffic dataset with real vehicle trajectories. A new four-dimensional dataset containing the amount and types of wireless data traffic, as well as the vehicle coordinates at each timestamp, can be obtained through the proposed approach. With this dataset in hand, we further design a flexible and embeddable spatiotemporal partial (STeP) convolutional network module to accurately model the traffic data of the IoV with a time series relationship. Moreover, the proposed STeP module is embedded into UNet [16] to conduct spatiotemporal traffic prediction. The experimental results confirm the promising performance of the proposed method.

2.1 Dataset Description and Processing

2.1.1 Dataset Merging

As mentioned above, although datasets tracking the trajectories of vehicles [17] and communication behaviors [18] separately exist, few works have merged them for conducting wireless data traffic prediction in the context of the IoV. As a result, the first mission in this work is to match a vehicle location information dataset with a communication traffic dataset. Denoting these two datasets as V and U, respectively, they can both be expressed as three-dimensional tensors. Specifically, $V = (N, T, P)$ records all the location information of N vehicles in T time stamps, where P contains the location information of each vehicle (such as its latitude and longitude information). On the other hand, the communication traffic dataset $U = (N', T', S)$ provides S types of communication traffic for N' users in T' time stamps. The main focus here is to match S with P, where the relationships between N and N', as well as T and T', should first be established as:

$$N = \Psi(\mathrm{N}'), T = \Phi(\Psi(T')), \tag{2.1}$$

where the $\Psi(\cdot)$ function gives the correspondence between the vehicles and the users. The intuition here is that we can treat the N users in U as passengers scattered across N' vehicles. $\Phi(\cdot)$ is the scaling function, which forces T to be on the same time scale as that of T'. With the above two equations in hand, the matching of these two datasets can be formulated as follows:

$$\begin{aligned} MS &= (\Psi(\Phi(N')), \Psi(\Phi(T')), P, \Psi(\Phi(S))) \\ &= (N, T, P, \Psi(\Phi(S))). \end{aligned} \tag{2.2}$$

To better illustrate the above process, Fig. 2.1 shows more details about the process of mapping the user communication data to the vehicle trajectory data. Note that for ease of understanding, here, we consider the special case in which $N = N'$. In such a case, $\Phi(\cdot)$ degrades to the identity mapping function to bind each vehicle with one specific user, where the communication behavior of each user during the movement of the vehicle can be readily obtained.

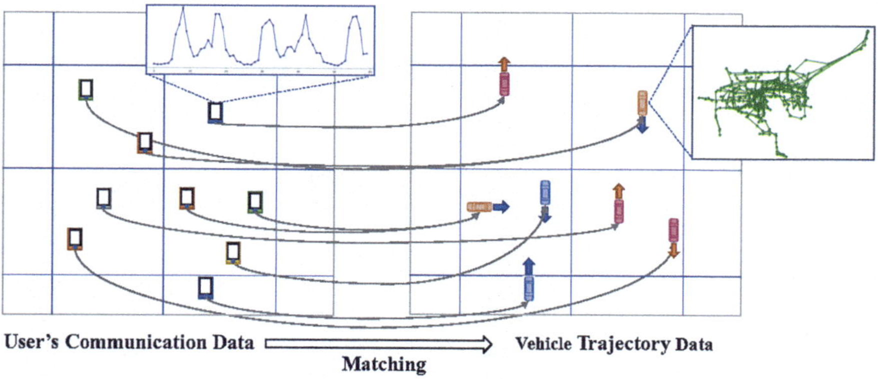

Fig. 2.1 Communication data are used to match the vehicle trajectory data

2.1.2 Traffic Map Division and Data Processing

After obtaining the merged dataset MS, we focus on tracking the dynamics of wireless data traffic on the basis of the vehicle trajectories at the city-wide scale. Specifically, in this work, we take taxi trajectories collected in Beijing as the basis dataset [14], from which the maximum and minimum latitudes/longitudes can be obtained to determine the moving areas of vehicles. As illustrated in Fig. 2.2, we first divide the corresponding citywide map into $H \times W$ grids, with the height and width of each grid denoted as G_H and G_W, respectively. Then, in each grid, the amount of wireless traffic that occurs during the tth time slot is calculated independently for different communication types by aggregating the corresponding wireless flows from the vehicles that are currently positioned within that grid. This aggregation operation is illustrated in the following definition.

Definition 2.1 Let P be a collection of trajectories in the tth time slot. For a Grid(i,j) that lies in the ith row and the jth column, the volume of the traffic in time slot t is defined as follows:

$$x_t^{i,j} = \sum\nolimits_{\mathrm{Tr} \in \mathrm{P}} |\{k \geq 1 | g_{k-1} \notin (i,j) \wedge g_k \in (i,j)\}| \tag{2.3}$$

where $\mathrm{Tr} : g_1 \rightarrow g_2 \rightarrow \cdots \rightarrow g_{\mathrm{Tr}}$ is the trajectory from P. g_k denotes the coordinates of one target point in the city-wide map. $|\cdot|$ is defined as the cardinality of the trajectory set. By performing such an aggregation operation in each grid per time slot, a set of wireless traffic-related tensors $\{x_t^{i,j}\}$ with dimensions of $T \times H \times W$ can be obtained. The entire process is described in Algorithm 2.1, where $\{\mathrm{LT}_{\max}, \mathrm{LT}_{\min}\}$ and $\{\mathrm{LN}_{\max}, \mathrm{LN}_{\min}\}$ represent the maximum/minimum latitudes and longitudes of the city, respectively.

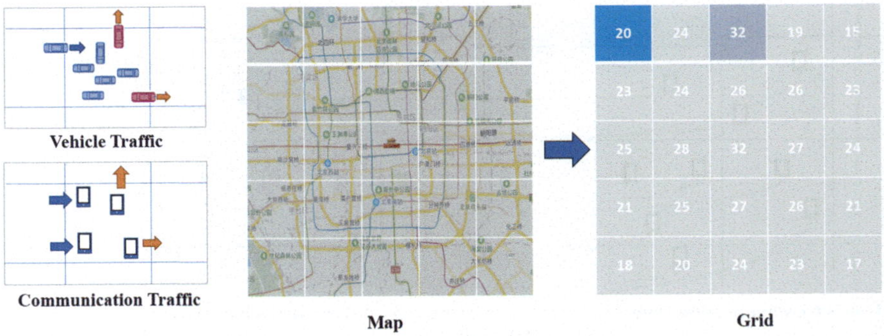

Fig. 2.2 Dividing the city map and generating a matrix

Algorithm 2.1: Matching the Vehicle Traffic Data and Communication Traffic Data

Input: V, U, H, W

Compute T, S, G_H, G_W, LT_{max}, LT_{min}, LN_{max}, LN_{min} according to $V$and U;

Initial $Grid = Zero(shape = [T, X, Y, S])$.

For $i = 1 \rightarrow T$ **do**

$Car_{pos} = V^i$;

For $j = 1 \rightarrow$ N **do**

$Map_{pos} = Car^j_{pos}$;

If then $LT_{max} > Map_{pos} \geq LT_{\min}$ **or** $LN_{max} > Map_{pos} \geq LN_{\min}$

$$P_{LT} = \frac{Map^{LT}_{pos} - LT_{min}}{G_H};$$

$$P_{LN} = \frac{Map^{LN}_{pos} - LN_{min}}{G_W};$$

$Grid[i, P_{LT}, P_{LN}]_{pos} + = V[j, i]$;

$Grid[i, P_{LT}, P_{LN}]_U + = \Phi(\Psi(U[j, i]))$;

end If

end For

end For

Through Algorithm 2.1, the time series of vehicle traffic and its counterpart for wireless data traffic can be obtained, where the former is a 4-dimensional tensor that illustrates the spatiotemporal distribution of the different types of wireless traffic volumes. The fundamental network architecture can be designed to model and predict traffic patterns.

2.2 Motion and Communication Behavior Prediction for Vehicles

2.2.1 Review of the ResNet and DenseNet Modules

To better understand the proposed STeP module, we quickly explore the basic ideas of ResNet and DenseNet. The features of the input and the lth layer of the network are denoted as X_0 and X_l, respectively. The structure of ResNet can be defined as follows:

$$X_l = X_l + f(X_{l-1}), \tag{2.4}$$

where $f(\cdot)$ is defined as a composite function of multiple convolutional layers, which is containing several consecutive operations such as convolution, batch normalization and a rectified linear unit [19–21]. X_l in Eq. (2.4) is updated recursively, in which all the features are treated as units through $f(\cdot)$ and then aggregated together through summation.

The basic structure of DenseNet is expressed as follows:

$$X_l = [X_{l-1}, f(X_{l-1})], \tag{2.5}$$

where $[\cdot, \cdot]$ represents the concatenation operation conducted on the feature set. Eq. (2.5) is quite similar to Eq. (2.4), except that the feature summation operation is replaced by concatenation. Both networks inherit the advantages of obtaining efficient gradients during backpropagation by fusing the features of the front and back layers. However, these methods also have drawbacks. For example, ResNet adds all the front-layer features to the back layer and hence has more parameters. DenseNet merges only the front-layer features but does not fuse them. This may reduce the complexity of the network but decrease the resulting performance.

To inherit the advantages of both networks and make them suitable for traffic prediction tasks, a STeP module is proposed in this paper. The basic idea is to divide the features contained in each layer into multiple groups and appropriately apply summation or concatenation operations in different groups. In this way, a light-weight variant network incorporating ResNet and DenseNet can be constructed. The details of such a network module are described in the following subsection.

2.2.2 STeP Module

In this subsection, we present the framework of the proposed STeP module. As mentioned above, we can divide the feature channels of a deep learning network into G groups. Denoting the features in the jth$(1 \leq j \leq G)$ group of layer l as g_l^j, the features of layer l can be defined as:

$$X_l = \left[g_l^1, \cdots, g_l^G\right], \tag{2.6}$$

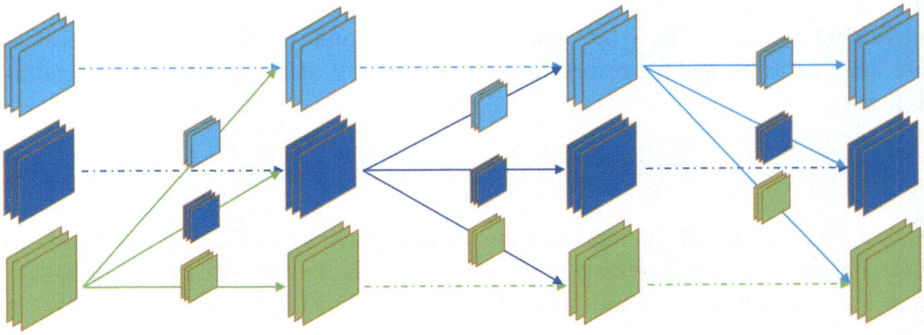

Fig. 2.3 The structure of the STeP module

Consequently, the block structure of the STeP module can be defined as:

$$X_l = X_{l-1} + f\left(g_{l-1}^{j-1}\right). \tag{2.7}$$

Figure 2.3 illustrates the structure of a single layer of the STeP module, which is divided into 3 groups. The number of features in each traffic pattern is amplified via group convolution to obtain G group features $\left[g_l^1, \cdots, g_l^G\right]$. All group features can then be regarded as the input X_0 in the STeP module. The purpose of dividing the features into multiple groups is to capture the different traffic patterns contained in the IoV, with each group convolution used to model one corresponding pattern. The STeP module makes full use of the advantages of ResNet and DenseNet so that the gradient can be efficiently backpropagated. Moreover, with the proposed module, one can independently apply summation or concatenation operations for different groups according to the different types of patterns they address. This process helps reduce the number of parameters required by the whole model and can potentially reach a good tradeoff between complexity and performance. Notably, each group represents only one of multiple traffic patterns. The STeP module can still sequentially fuse and learn each traffic pattern as the layers become deeper. This finding reveals that the STeP module can effectively model traffic data with spatiotemporal relationships.

2.2.3 STeP-UNet

In this subsection, we consider embedding the STeP module into the UNet architecture to construct the final learning model for prediction purposes. Figure 2.4

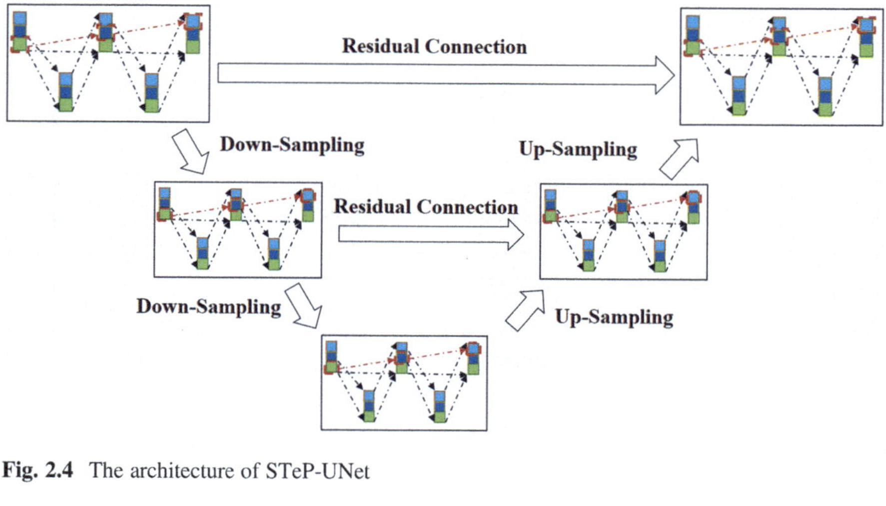

Fig. 2.4 The architecture of STeP-UNet

illustrates the architecture of the proposed STeP-UNet, where each node in UNet is affected by the STeP module. STeP-UNet has many skip connections from the architecture level to the node level. This characteristic enables the model to efficiently learn the timing relationships of traffic data and thereby avoid the problem of gradient dispersion.

Regarding the parameter settings of the model, summation operations are used for overlapping channels, and concatenation operations are used for redundant channels. These operations can reduce the number of required network parameters, which can be analyzed from two aspects. On the one hand, the STeP module sequentially uses 3×3 and 1×1 convolutions in its amplification function. The number of group channels is amplified to $\frac{C}{4}$ via 3×3 convolution, and then, it is amplified to C via 1×1 convolution. The number of parameters in the 1×1 convolution is much less than that of the 3×3 convolution. In other words, the number of network parameters decreases more as the number of groups increases than do those of other networks. On the other hand, unlike STResNet and STDenseNet, which use multiple ResNet or DenseNet modules to separately capture multiple traffic patterns, the STeP module employs a single module for multipattern data learning. This is because the STeP module uses group convolution to separately model multiple patterns, where each mode corresponds to only one group and each group sequentially performs partial-to-whole feature fusion. On the basis of these characteristics, STeP-UNet can greatly reduce the number of required network parameters and speed up the network training process. Notably, the STeP module can also be embedded in any network architecture to achieve efficient feature integration. The purpose of choosing the UNet architecture in this work is to achieve direct skip connections from the front layer to the back layer.

2.3 Simulations and Discussions

2.3.1 *Datasets*

We use datasets acquired from two different domains to model the moving trends of vehicles and the communication traffic of users. The vehicle traffic dataset is obtained from taxi information in Beijing. The end-user communication dataset is derived from user communication data in Milan, Italy, and the BS traffic data of a city in China. To capture both the temporal and spatial correlations of the datasets, the samples are collected via a sliding window in the temporal dimension. The details of the corresponding datasets are described below.

Vehicle Traffic: This dataset contains the GPS trajectories of 10,357 taxis within Beijing from 02/02/2008 to 02/08/2008 [17]. The total number of points included in this dataset is approximately 15 million, and the total distance of the trajectories reaches nine million kilometers.

Cellular Traffic: Inspired by [1], we employ the cellular traffic dataset of [18], which is provided by Telecom Italia. The dataset was collected from 11/01/2013 to 01/01/2014 with a temporal interval of 10 min across the entire city of Milan (62 days, 300 million records, approximately 19 GB). Three kinds of cellular traffic were recorded by the service provider: short message service (SMS; the incoming and outgoing SMS directions are denoted as In-SMS and Out-SMS, respectively), call service (incoming and outgoing calls are denoted as In-Call and Out-Call, respectively), and internet service traffic.

BS Traffic: Real communication traffic data, including uplink and downlink communication traffic data, were collected from all BSs in a certain city in China. The dataset was collected from 08/06/2019 to 11/07/2019 across the whole city with a temporal interval of 60 min (94 days, 51.9 million records, approximately 17.1 GB).

The traffic data are composed of temporal closeness, period, and trend information as their three major modeling components. We first transform the traffic data throughout the city at each time slot into a multichannel image-like tensor via the approach introduced in Definition 2.1. The 6-channel vehicle-cellular traffic tensors (Vehicle Traffic, In-SMS, Out-SMS, In-Call, Out-Call, and Internet) and 3-channel vehicle-BS traffic tensors (Vehicle Traffic, Up-Link, and Down-Link) are separately fed into the first three components of the network to model the three temporal properties: closeness, period, and trend. However, in STeP-UNet, the first three network components share the same network structure with a group convolutional layer followed by a STeP module sequence. Finally, the aggregated features are mapped to $[0, 1]$ via a sigmoid function $S(x) = \frac{1}{1+e^{-x}}$, which yields faster convergence than does the standard logistic function during the process of backpropagation learning.

2.3.2 Settings

Dataset Settings For the traffic matching process, the whole city is split into 32×32 grids, and min–max normalization is used to convert the whole traffic volume to the $[0, 1]$ scale. For functions Ψ and Φ, the truncation operation is first performed, and then identity mapping is used to match the data. Finally, each vehicle trajectory corresponds to a unique communication process. During the prediction stage, the prediction values are denormalized and used for evaluation purposes. Note that to preserve not only the temporal correlations but also the spatial correlations in the original dataset, the sliding window method is employed when conducting sampling. Eighty percent of the samples are randomly chosen for training, while the remaining samples are left for testing [21].

Loss Function The proposed deep learning model can be easily trained by minimizing the Frobenius norm between the predicted value and the ground-truth value of the tth slot:

$$L(\theta) = \text{argmin}_{\theta}\left|\left|X_t - \tilde{X}_t\right|\right|_F^2, \tag{2.8}$$

where X_t is the ground-truth traffic map and $\tilde{X}_t$ is the predicted traffic map in time slot t.

Hyperparameter Settings STeP-UNet is trained using one NVIDIA Tesla V100 GPU under the PyTorch framework. The well-known adaptive moment estimation (Adam) optimization method is employed for gradient descent, given that the initial learning rate is 0.001 [22]. We choose the hyperparameters according to those of the optimal model determined on the validation set.

Network Settings For the proposed STeP-UNet, we set the initial number of convolution channels to 4G, where G denotes the number of traffic types, which equals 6 in this work. Moreover, to conduct a performance comparison, we also examine two other competitive schemes, namely, STResNet and STDenseNet, with similar network settings to those listed in [21]. More details can be found therein.

2.3.3 *Performance*

We compared the performances achieved by STeP-UNet, STResNet, and STDenseNet on the vehicle-cellular and vehicle-BS datasets. The results are presented in Table 2.1. For different indicators, such as the number of parameters (Params), the mean absolute error (MAE), the root mean square error (RMSE), and the coefficient of determination (R2), STeP-UNet demonstrates good generalization performance.

On the vehicle-cellular dataset, STeP-UNet achieves the lowest MAE and RMSE values relative to those of STResNet and STDenseNet. In terms of the MAE, the performance of STeP-UNet is significantly improved. In terms of the RMSE metric, STePUNet achieves 4% and 8% performance improvements over STResNet and

Table 2.1 Traffic prediction performance comparison among STeP-UNet, STResNet and STDenseNet

Dataset	Model	Params (M)	MAE	RMSE	R2
Vehicle-cellular	STResNet	0.85	0.09	0.24	72.54
	STDenseNet	0.78	0.10	0.25	72.05
	STeP-UNet	0.75	**0.07**	**0.23**	**75.85**
Vehicle-BS	STResNet	0.85	0.56	1.56	94.50
	STDenseNet	0.78	0.87	2.19	88.76
	STeP-UNet	0.75	**0.55**	**1.55**	**94.52**

STDenseNet, respectively. The proposed STeP-UNet also performs best on the vehicle-BS dataset. The reason for this is that the STeP module allows the network to more effectively capture the information of the temporal and spatial domains, especially when the feature size is large. Moreover, the UNet structure facilitates the sharing of parameters among multiple traffic patterns, which again confirms the promising performance of the proposed model in the task of jointly predicting both communication and movement behaviors.

2.3.4 *Visualization and Analysis*

Figures 2.5 and 2.6 depict the vehicle traffic and communication traffic predictions from the perspectives of the spatial and temporal domains, respectively. Figure 2.5 shows that the vehicle traffic and its communication traffic pattern can be well matched. It generally obeys the rule that communication traffic increases with vehicle traffic, but exceptions may occur in some areas. For example, consider the city center. Figure 2.6 also shows the diversity of different traffic patterns, but they are all positively correlated with vehicle traffic.

Figure 2.6 shows the time-varying relationships among various traffic patterns. Note that different types of traffic have different magnitudes, but they have almost the same timing relationships. As shown in the figure, although some local disturbances are present, STeP-UNet can better predict traffic patterns in the IoV scenario.

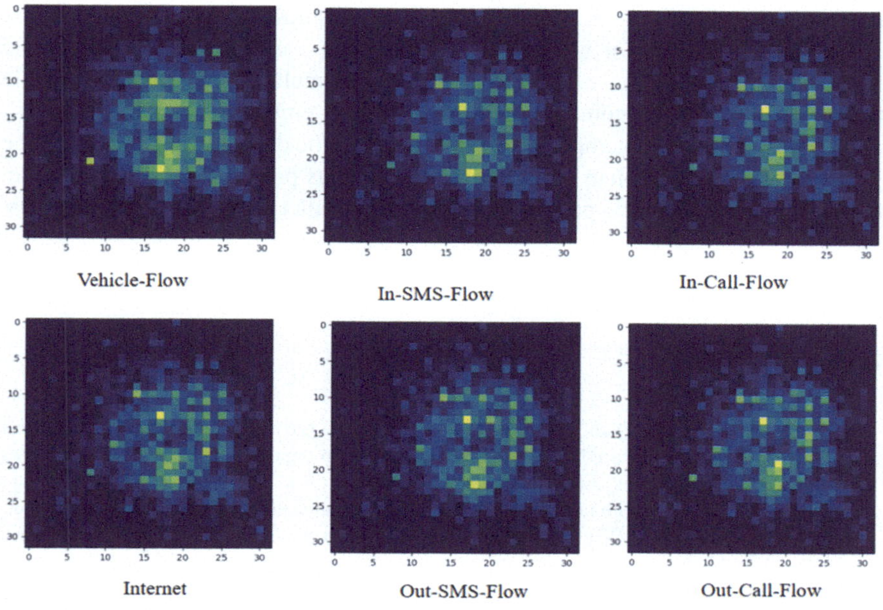

Fig. 2.5 Spatial patterns of the vehicle-cellular traffic prediction results

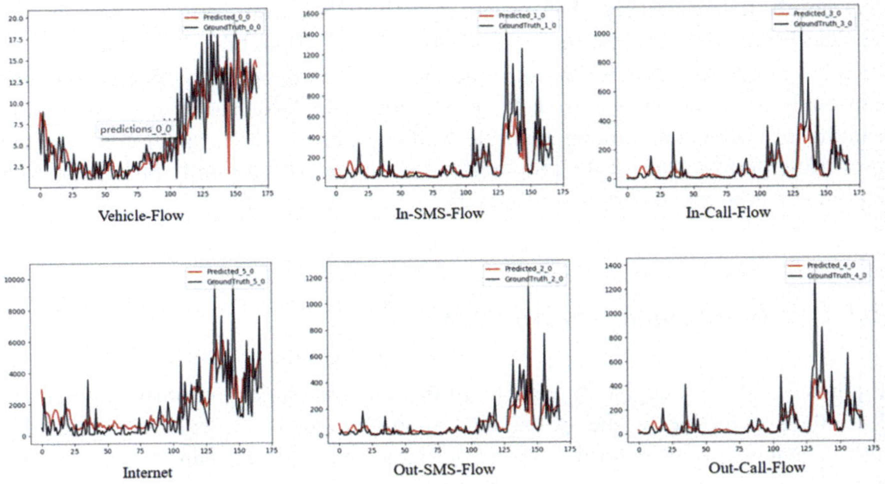

Fig. 2.6 Temporal patterns of the vehicle-cellular traffic prediction results

2.4 Summary

In this chapter, we studied traffic prediction in the IoV scenario. To collect data from an IoV scenario, a traffic matching algorithm was proposed to match vehicle traffic with real communication traffic, where vehicle traffic was statistically analyzed and modeled on a city-wide scale. To better predict various traffic patterns, an STeP module, which is based on spatiotemporal relationships, was used. This module models each traffic pattern as a group, and each group can perform partial-to-whole feature fusion through an expansion operation to simultaneously predict multiple traffic patterns. Furthermore, the STeP module was embedded into UNet to efficiently capture the spatiotemporal relationships of traffic data, thereby alleviating the gradient dispersion problem. The experimental results proved that STeP-UNet can capture a variety of traffic patterns and precisely predict traffic volumes in an IoV scenario.

References

1. Zhang, C., Zhang, H., Qiao, J., Yuan, D., & Zhang, M. (2019). Deep transfer learning for intelligent cellular traffic prediction based on cross-domain big data. *IEEE Journal on Selected Areas in Communications, 37*(6), 1389–1401.
2. Fan, W., et al. (2023). Joint task offloading and resource allocation for vehicular edge computing based on V2I and V2V modes. *IEEE Transactions on Intelligent Transportation Systems, 24*(4), 4277–4292.
3. Noor-A-Rahim, M., Liu, Z., Lee, H., Ali, G. G. M. N., Pesch, D., & Xiao, P. (2022). A survey on resource allocation in vehicular networks. *IEEE Transactions on Intelligent Transportation Systems, 23*(2), 701–721.

4. Kuutti, S., Bowden, R., Jin, Y., Barber, P., & Fallah, S. (2021). A survey of deep learning applications to autonomous vehicle control. *IEEE Transactions on Intelligent Transportation Systems, 22*(2), 712–733.
5. Mozaffari, S., Al-Jarrah, O. Y., Dianati, M., Jennings, P., & Mouzakitis, A. (2022). Deep learning-based vehicle behavior prediction for autonomous driving applications: A review. *IEEE Transactions on Intelligent Transportation Systems, 23*(1), 33–47.
6. Zhu, X., Luo, Y., Liu, A., Tang, W., & Bhuiyan, M. Z. A. (2021). A deep learning-based Mobile Crowdsensing scheme by predicting vehicle mobility. *IEEE Transactions on Intelligent Transportation Systems, 22*(7), 4648–4659.
7. Qiao, Y., Yin, J., Wang, W., Duarte, F., Yang, J., & Ratti, C. (2023). Survey of deep learning for autonomous surface vehicles in marine environments. *IEEE Transactions on Intelligent Transportation Systems, 24*(4), 3678–3701.
8. Wang, Z., & Wong, V. W. S. Bayesian meta-learning for adaptive traffic prediction in wireless networks. *IEEE Transactions on Mobile Computing, Early Access*. https://doi.org/10.1109/TMC.2023.3325301d
9. Shen, W., Zhang, H., Guo, S., & Zhang, C. (2021). Time-wise attention aided convolutional neural network for data-driven cellular traffic prediction. *IEEE Wireless Communications Letters, 10*(8), 1747–1751.
10. George, S., & Santra, A. K. (2020). Traffic prediction using multifaceted techniques: A survey. *Wireless Personal Communications, 115*, 1047–1106.
11. Nan, H., Li, R., Zhu, X., Ma, J., & Xue, K. Spatio-temporal identity multi-graph convolutional network for traffic prediction in the metaverse. *IEEE Journal on Selected Areas in Communications, Early Access*. https://doi.org/10.1109/JSAC.2023.3345389
12. Zhao, N., Wu, A., Pei, Y., Liang, Y.-C., & Niyato, D. (2022). Spatial-temporal aggregation graph convolution network for efficient Mobile cellular traffic prediction. *IEEE Communications Letters, 26*(3), 587–591.
13. Zhang, C., Zhang, H., Yuan, D., & Zhang, M. (2018). Citywide cellular traffic prediction based on densely connected convolutional neural networks. *IEEE Communications Letters, 22*(8), 1656–1659.
14. Lin, Z., Feng, J., Lu, Z., Li, Y., & Jin, D. (2019). Deepstn+: Context-aware spatial-temporal neural network for crowd flow prediction in metropolis. In *Proceeding of the thirty-first AAAI conference on artificial intelligence (AAAI-17)*.
15. Zhang, Q., Yu, K., et al. (2022). Graph neural network-driven traffic forecasting for the connected internet of vehicles. *IEEE Transactions on Network Science and Engineering, 9*(5), 3015–3027.
16. Ronneberger, O., Fischer, P., & Brox, T. (2015). U-net: Convolutional networks for biomedical image segmentation. In *Medical Image Computing and Computer- Assisted Intervention (MICCAI), series LNCS* (Vol. 9351, pp. 234–241). Springer.
17. Yuan, J., Zheng, Y., Xie, X., & Sun, G. (2011). Driving with knowledge from the physical world. In *Proceedings of the 17th ACM SIGKDD international conference on knowledge discovery and data mining* (pp. 316–324).
18. Barlacchi, G., Barlacchi, G., Larcher, R., & Casella, A. (2015). A multi-source dataset of urban life in the city of Milan and the province of trentino. *Scientific Data, 2*, 2052–4463.
19. Ioffe, S., & Szegedy, C. (2015). Batch normalization: accelerating deep network training by reducing internal covariate shift. In *ICML* (pp. 448–456).
20. Glorot, X., Bordes, A., & Bengio, Y. (2012). Deep sparse rectifier neural networks. In *International conference on artificial intelligence and statistics*.
21. Zhai, L., Yang, Y., Song, S., Ma, S., Zhu, X., & Yang, F. (2021). Selfsupervision spatiotemporal part-whole convolutional neural network for traffic prediction. *Physica A: Statistical Mechanics and its Applications, 579*, 126141.
22. Kingma, D. P., & Ba, J. (2014). Adam: A method for stochastic optimization. *arXiv preprint arXiv:1412.6980*.

Chapter 3
Social Mobility-Aware Communication Resource Management

In the existing works on underlaid C-V2X networks with high sum rates, several challenging issues remain to be addressed. First, the existing schemes heavily rely on instantaneous information such as location and CSI information, which results in heavy signaling overhead and system latency levels due to the frequent resource reallocation and vehicle repairing steps caused by rapid vehicle movements. To address this issue, Abbas et al. [1] incorporated vehicle speed, intervehicle distance, and link lifetime information into a vehicle pairing strategy and proposed a novel approach for alternatively optimizing the spectrum assignment and power control tasks. However, the work in [1] did not consider social relationships. Second, the existing works have focused on individually optimizing vehicle pairings and radio resources. Owing to the highly coupled nature of vehicle pairing and radio resources, jointly optimizing these variables could further increase the sum rate. Finally, Hu, Zhang, and Abbas et al. [1–3] mainly focused on vehicle pairing and radio resource optimization by assuming that systems work in the overlaid mode. In this mode, vehicular and cellular users use different spectrum resources. This type of approach leads to low spectrum utilization. To improve it, the best strategy is to adopt the underlaid mode, but this technique leads to severe cochannel interference, which in turn results in a low system sum rate.

To address these three problems and improve the sum rate and spectrum utilization level of the employed system, this work establishes a social mobility-aware underlaid C-V2X framework and designs a joint vehicle pairing and radio resource optimization strategy. The contributions of this paper can be summarized as follows.

- We design a social mobility-aware underlaid C-V2X transmission framework, where multifarious service information can be simultaneously transmitted in a nonorthogonal manner. In the framework, social properties (social communities and social ties) and physical characteristics (distances and velocity differences) are accounted for. An optimization problem is formulated to maximize the system sum rate by jointly determining the vehicle pairing, spectrum, and power

H. Zhang et al., *Intelligent Resources Management for Vehicular Social Networks*, Wireless Networks, https://doi.org/10.1007/978-3-031-80169-3_3

resources, and this is proven to be a nondeterministic polynomial-time (NP)-hard problem.

- A joint vehicle pairing, spectrum assignment, and power control algorithm is proposed to solve the NP-hard problem. More specifically, the sum-rate maximization problem is decoupled into two separate subproblems, a joint vehicle pairing and spectrum assignment subproblem and a power control subproblem, which are solved in an iterative manner.
- To address the first subproblem, we design a heuristic social mobility-aware vehicle pairing algorithm (HSMA-VPA) and a Kuhn-Meyer-based spectrum assignment algorithm (KM-SAA). With a given vehicle pairing and spectrum assignment strategy, we propose a low-complexity three-dimensional geometric power control algorithm (3D-PCA) to determine the transmission power.
- We conduct intensive simulations to evaluate the sum-rate performance of the proposed algorithms. The simulation results show that the NOMA-enabled scheme clearly outperforms the traditional orthogonal multiple-access mechanism. Additionally, the superiority of utilizing social attributes is verified. Moreover, compared with the baselines, the proposed algorithms result in a significant performance gain in terms of the sum rates produced for different system setups.

The remainder of this chapter is organized as follows. Section 3.1 describes the social domain and physical domain underlying the C-V2X network framework. An optimization problem is then formulated to maximize the sum rate of the underlaid C-V2X network while guaranteeing the minimum QoS requirements of differentiated V2X users and cellular users in Sect. 3.2. Section 3.3 proposes a NOMA-JVP-SA-PCA. The simulation results are presented in Sect. 3.4. Finally, Sect. 3.5 concludes the entire chapter.

3.1 System Model

This work studies a NOMA-based underlaid C-V2X network to ensure that the differentiated QoS requirements of each V2X service and cellular service are satisfied, as shown in Fig. 3.1. In this network, a continuous period is divided into time slots with fixed durations denoted by $t \in \mathcal{T} = \{1, 2, \ldots, T\}$, a macro base station (MBS) is deployed to support the high-uplink-rate data transmissions of U cellular users indexed by $u \in \mathcal{U} = \{1, 2, \ldots, U\}$, and S roadside units (RSUs) indexed by $s \in \mathcal{S} = \{1, 2, \ldots, S\}$ are deployed to support diversified V2X services. In this scenario, M vehicular users indexed by $m \in \mathcal{M} = \{1, 2, \ldots, M\}$ can communicate with the RSUs (i.e., vehicle-to-infrastructure (V2I) links), which provides a high rate of required services. Moreover, the M vehicular users can also communicate with their paired vehicular users (i.e., vehicle-to-vehicle (V2V) links), thus supporting services with high reliability requirements. In this system, the alternative V2V receiving user set is denoted by $v \in \mathcal{V} = \{1, 2, \ldots, V\}$. In Fig. 3.1, the system model is divided into two domains: the physical domain and

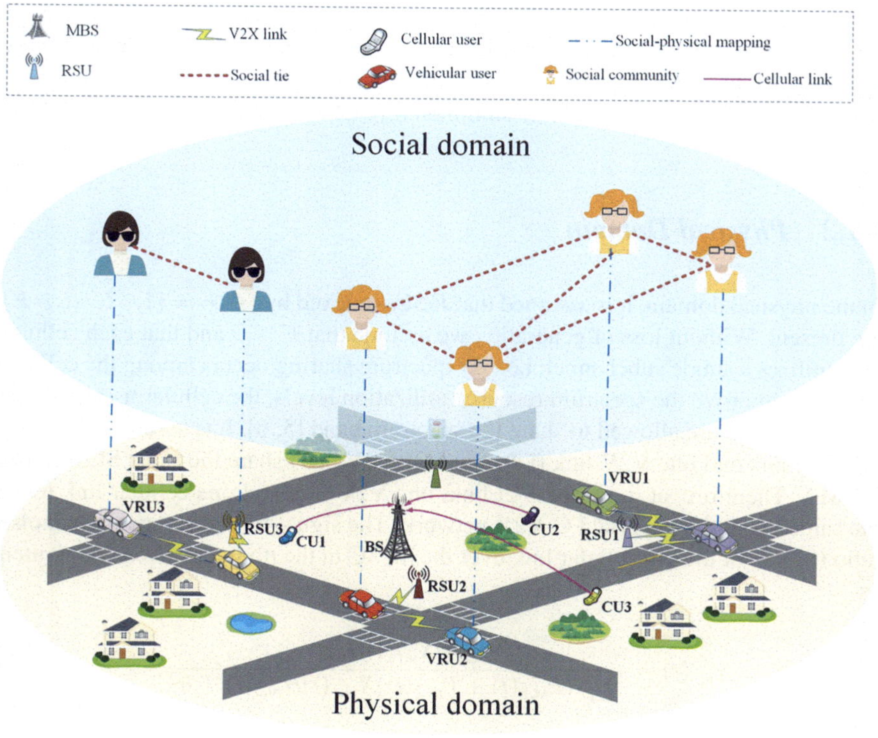

Fig. 3.1 System model for social mobility-aware underlaid C-V2X networks

the social domain. In the social domain, the social behaviors of users represent their social connections, which can be obtained from social media such as Facebook. The establishment of communication links in the physical domain is mainly determined by the distances between users.

3.1.1 Social Domain

According to the concept of social networks [4], social characteristics depend on the real activities and behaviors of corresponding human beings, which have two main characteristics: social communities and social ties. A social community refers to a group of users who share common interests in specific information or content, such as destinations, traffic conditions, road conditions, and weather. The social ties in a social network represent the relationships or levels of trust among users. Leveraging social communities and social ties can help build trustworthy V2V links and improve the sum rate of the utilized system.

In an underlaid C-V2X network, it is assumed that A social communities indexed by $a \in \mathcal{A} = \{1, 2, \ldots, A\}$ are present. The social ties are indexed by $\boldsymbol{\delta} = \{\delta_{m,v}(t) \mid \delta_{m,v}(t) \in \{0, 1\}\}$. If the mth vehicle shares its content of interest[1] with the vth vehicle in the tth time slot, then $\delta_{m,v}(t) = 1$; otherwise, $\delta_{m,v}(t) = 0$.

3.1.2 *Physical Domain*

In the physical domain, it is assumed that F SCs indexed by $f \in \mathcal{F} = \{1, 2, \ldots, F\}$ are present. Without loss of generality, we assume that $F = U$ and that each cellular user utilizes a single subchannel; i.e., no spectrum sharing occurs among the cellular users. To improve the spectrum resource utilization levels, the cellular users and the V2I-V2V pairs are allowed to share the spectrum band [5, 6]. It is assumed that only one V2I link and one V2V link (i.e., one V2I-V2V pair) share the same SC through NOMA. Therefore, at most one V2I link, one V2V link and one cellular link reuse the same SC in the underlaid C-V2X network. The signal-to-interference-plus-noise ratio (SINR) of the uth cellular link over the fth SC in the tth time slot can be written as

$$\gamma_{u,f}(t) = \frac{P_u(t) X_{u,f}(t) H_u(t)}{\sum\limits_{m \in \mathcal{M}} (P_{m,1}(t) + P_{m,2}(t)) X_{m,f}(t) H_{m,u}(t) + \sigma^2}, \tag{3.1}$$

Where, $P_{m,1}(t)$, $P_{m,2}(t)$, and $P_u(t)$ denote the transmission powers of the mth V2I link, the mth V2V link and the uth cellular link over the fth SC in the tth time slot, respectively. σ^2 is the noise power. A binary variable $X_{m,f}(t) \in \{0, 1\}$ is defined as the spectrum assignment factor. If the mth vehicular user occupies the fth SC in the tth time slot, $X_{m,f}(t) = 1$; otherwise, $X_{m,f}(t) = 0$. Similarly, $X_{u,f}(t)$ is also a spectrum assignment indicator. $H_u(t)$ is the channel power gain of the uth cellular user in the tth time slot, which is denoted as

$$H_u(t) = \beta_u(t) \cdot h_u(t) \cdot \phi_u \cdot (d_u(t))^{-\alpha_u}, \tag{3.2}$$

in which $d_u(t)$ is the distance between the uth cellular user and the MBS, α_u and ϕ_u represent the path loss exponent and constant, respectively, $h_u(t)$ represents the shadowing fading factor that is assumed to be a log-normal random variable, and $\beta_u(t)$ is the small-scale fading factor. Similarly, $H_{m,u}(t)$ represents the interference channel power gain from the mth vehicle user who transmits data to the RSU and from another vehicle user to the uth cellular user.

[1]This includes the required high-rate information transmitted through V2I links and the required high-reliability information transmitted through V2V links.

The SINRs at the V2I receiver and the V2V receiver over the fth SC in the tth time slot are expressed as

$$\gamma_{m,1,f}(t)=\frac{P_{m,1}(t)X_{m,f}(t)H_{m,s}(t)}{\sum\limits_{u\in\mathcal{U}}P_u(t)X_{u,f}(t)H_{u,m,1}(t)+I_{in}^{V2I}(t)+\sigma^2},\tag{3.3}$$

$$\gamma_{m,v,2,f}(t)=\frac{P_{m,2}(t)X_{m,f}(t)\Psi_{m,v}(t)H_{m,v}(t)}{\sum\limits_{u\in\mathcal{U}}P_u(t)X_{u,f}(t)\Psi_{m,v}(t)H_{u,m,2}(t)+I_{in}^{V2V}(t)+\sigma^2},\tag{3.4}$$

where $I_{in}^{V2I}(t)=P_{m,2}(t)X_{m,f}(t)H_{m,s}(t)$ and $I_{in}^{V2V}(t)=P_{m,1}(t)X_{m,f}(t)H_{m,v}(t)$. $H_{m,\ s}(t)$ and $H_{m,\ v}(t)$ are the channel power gains of the mth V2I link and the mth V2V link, respectively; $H_{u,\ m,\ 1}(t)$ denotes the interference channel power gain from the uth cellular user to the mth V2I link; and $H_{u,\ m,\ 2}(t)$ is the interference channel power gain from the uth cellular user to the mth V2V link. By combining the physical and social relationships, the vehicle pairing variable in the tth time slot is denoted as

$$\Psi_{m,v}(t)=k_{m,v}(t)\cdot\delta_{m,v}(t).\tag{3.5}$$

where $k_{m,\ v}(t)$ is the physical relationship between the mth vehicle and the vth vehicle, which can be defined as follows:

$$k_{m,v}(t)=\begin{cases}1, & \text{if } \beta>\rho_{m,v}(t),\\ 0, & \text{otherwise.}\end{cases}\tag{3.6}$$

where β represents the threshold of the physical domain. $\rho_{m,\ v}(t)$ is written as

$$\rho_{m,v}(t)=\alpha\cdot f(|d_m(t)-d_v(t)|)+(1-\alpha)\cdot f(|v_m(t)-v_v(t)|),\forall m\in\mathcal{M},v\in\mathcal{V},\tag{3.7}$$

where $|d_m(t)\ -\ d_v(t)|$ is the distance between the mth vehicle and the vth vehicle, $\alpha\in[0,\ \ 1]$ represents the weight of the relative distance, $|v_m(t)-v_v(t)|$ is the velocity difference, and $f(\cdot)$ is the normalized function.

Based on the previous expressions of the SINR, the corresponding achievable data rates of the uth cellular link, the mth V2I link, and the mth V2V link (in the mth V2I-V2V pair) over the fth SC in the tth time slot can be sequentially expressed as shown below:

$$R_{u,f}(t)=B_{sc}\log_2(1+\gamma_{u,f}(t)),\tag{3.8}$$

$$R_{m,\ 1,f}(t)=B_{sc}\log_2(1+\gamma_{m,\ 1,f}(t)),\tag{3.9}$$

$$R_{m,v,2,f}(t)=B_{sc}\log_2(1+\gamma_{m,v,2,f}(t)),\tag{3.10}$$

where B_{sc} represents the bandwidth of each SC. According to (3.8)–(3.10), the system sum rate can be written as

$$R_{\text{sum}} = \sum_{t=1}^{T} \sum_{f=1}^{F} \sum\nolimits_{v=1}^{V} \sum\nolimits_{m=1}^{M} \sum\nolimits_{u=1}^{U} R_{\text{sum.sub}}(t) \ , \tag{3.11}$$

in which $R_{\text{sum}\_\text{sub}}(t) = R_{u,f}(t) + R_{m,1,f}(t) + R_{m,v,2,f}(t)$.

3.2 Spectrum Efficiency Maximization Problem Formulation

The system sum-rate performance of the described NOMA-based vehicular network is highly dependent on the vehicle pairing and resource (spectrum and power) management statuses among the V2V links, the V2I links and the cellular links. In this work, our objective is to maximize the system sum rate while guaranteeing the QoS requirements of differentiated users. To achieve this goal, the joint vehicle pairing, spectrum assignment, and power control optimization problems are formulated as

$$\mathcal{P}_1: \max_{\Psi_{m,v}(t), X_{m,f}(t), X_{u,f}(t), \text{atop} P_{m,1}(t), P_{m,2}(t), P_u(t)} R_{\text{sum}}, \tag{3.12a}$$

$$\text{s.t.} X_{m,f}(t), X_{u,f}(t) \in \{0, \ 1\}, \ \forall f \in \mathcal{F}, u \in \mathcal{U}, \forall \ m \in \mathcal{M}, t \in \mathcal{T} \tag{3.12b}$$

$$\Psi_{m,v}(t) \in \{0, \ 1\}, \forall m \in \mathcal{M}, v \in \mathcal{V}, t \in \mathcal{T}, \tag{3.12c}$$

$$\sum\nolimits_{u=1}^{U} X_{u,f}(t) = 1, \forall f \in \mathcal{F}, t \in \mathcal{T}, \tag{3.12d}$$

$$\sum\nolimits_{f=1}^{F} X_{u,f}(t) = 1, \ \forall \ u \in \mathcal{U}, t \in \mathcal{T}, \tag{3.12e}$$

$$\sum\nolimits_{m=1}^{M} X_{m,f}(t) = 1 \ , \forall \ f \in \mathcal{F}, t \in \mathcal{T}, \tag{3.12f}$$

$$\sum\nolimits_{f=1}^{F} X_{m,f}(t) \leq 1 \ , \forall \ m \in \mathcal{M}, t \in \mathcal{T}, \tag{3.12g}$$

$$R_{m, \ 1,f}(t) \geq R_{th, \ 1}^{m}, \forall \ m \in \mathcal{M}, f \in \mathcal{F}, t \in \mathcal{T}, \tag{3.12h}$$

$$R_{m,v, \ 2,f}(t) \geq R_{th, \ 2}^{m}, \forall \ m \in \mathcal{M}, v \in \mathcal{V}, \forall \ f \in \mathcal{F}, t \in \mathcal{T}, \tag{3.12i}$$

$$R_{u,f}(t) \geq R_{th}^{u}, \forall \ u \in \mathcal{U}, f \in \mathcal{F}, t \in \mathcal{T}, \tag{3.12j}$$

$$\frac{\sum\limits_{v \in \mathcal{V}} (P_{m,1}(t) - P_{m,2}(t)) \Psi_{m,v}(t) H_{m,2}(t)}{\sigma^2} \geq \xi, \forall m \in \mathcal{M}, t \in \mathcal{T}, \tag{3.12k}$$

$$R_{u,f}(t) \geq R_{th}^{u}, \forall\ u \in \mathcal{U}, f \in \mathcal{F}, t \in \mathcal{T}, \tag{3.12l}$$

$$0 \leq P_{m,1}(t) + P_{m,2}(t) \leq P_{m}^{\max}, \forall\ m \in \mathcal{M}, t \in \mathcal{T}, \tag{3.12m}$$

where $R_{th,1}^{m}$, $R_{th,2}^{m}$, and R_{th}^{u} are the minimum SINR thresholds for the mth V2I link, the mth V2V link, and the uth cellular link, respectively (the former two links are contained in the mth V2I-V2V pair); $P_{u}^{\max}$ and $P_{m}^{\max}$ denote the maximum transmission powers of the uth cellular user and the mth vehicular user, respectively; and ξ is the minimum power difference ratio, which depends on the utilized hardware [7]. Constraints (3.12b) and (3.12c) represent the fact that the SC assignment and vehicle pairing variables are binary. For the sake of fairness, (3.12d) and (3.12f) dictate that each SC can be reused by only one cellular user and one vehicular user. One user can only access one SC according to (3.12d) and (3.12g). Constraints (3.12h)-(3.12j) mean that the QoS requirements of different users should be satisfied. Constraint (3.12k) is the power difference condition that guarantees successful SIC at the receiver of each V2V link. Constraints (3.12l) and (3.12m) list the feasible transmission powers of the cellular users and vehicular users, respectively. The formulated sum-rate optimization problem $\mathcal{P}1$ is a mixed-integer nonlinear programming (MINLP) problem due to the hybrid binary variables $\Psi_{m,\ v}(t)$, $X_{m,\ f}(t)$, and $X_{u,\ f}(t)$ and continuous variables $P_{m,\ 1}(t)$, $P_{m,\ 2}(t)$, and $P_{u}(t)$. To solve this problem, the NOMA-JVP-SA-PCA is proposed. The detailed solution procedure is shown in the section below.

3.3 Graph Theory-Based Vehicle Association, Spectrum Assignment, and Power Control Strategy

As mentioned above, the formulated optimization problem $\mathcal{P}1$ is nonconvex, and it is difficult to obtain the global optimal solution. As an alternative, we propose converting the original problem into two subproblems: a binary variable optimization subproblem (i.e., a vehicle pairing and spectrum assignment subproblem $\mathcal{P}2$) and a continuous variable optimization subproblem (i.e., a power control subproblem $\mathcal{P}3$). $\mathcal{P}2$ and $\mathcal{P}3$ are given by

$$\begin{aligned} \mathcal{P}2: \quad & \max_{\Psi_{m,v}(t), X_{m,f}(t), X_{u,f}(t)} R_{\text{sum}}, \\ \text{s.t.} \quad & (3.12b) - (3.12g), \end{aligned} \tag{3.13}$$

$$\begin{aligned} \mathcal{P}3: \quad & \max_{P_{m,1}(t), P_{m,2}(t), P_{u}(t)} R_{\text{sum}}, \\ \text{s.t.} \quad & (3.12h) - (3.12i). \end{aligned} \tag{3.14}$$

By solving the first subproblem, the analytical expression of $\{\Psi_{m,\ v}(t)^{*}, X_{m,\ f}(t)^{*}, X_{u,\ f}(t)^{*}\}$ can be obtained. Similarly, $\{P_{m,\ 1}(t)^{*}, P_{m,\ 2}(t)^{*}, P_{u}(t)^{*}\}$ can be derived by solving the second subproblem. Then, by substituting the obtained power into the first subproblem, the vehicle pairing and spectrum assignment strategy is updated. Finally, $\{\Psi_{m,\ v}(t)^*, P_{m,\ 1}(t)^*, P_{m,\ 2}(t)^*, P_u(t)^*, X_{m,\ f}(t)^*, X_{u,\ f}(t)^*\}$ for the original problem can be obtained accordingly.

3.3.1 Vehicle Pairing and Spectrum Assignment

Owing to the nonconvex objective function and the binary variables $\Psi_{m,\ v}(t)$, $X_{m,\ f}(t)$, and $X_{u,\ f}(t)$, $\mathcal{P}2$ is still a nonconvex optimization subproblem. Generally, no efficient method is available for obtaining globally optimal solutions. To address this issue, this subsection further decouples $\mathcal{P}2$ into a vehicle pairing sub-subproblem and spectrum assignment sub-subproblem and iteratively solves them.

On the basis of the description provided in Sect. 3.2, HSMA-VPA is proposed, and it is described in Algorithm 3.1. To begin, the social tie set $\boldsymbol{\delta}$ is obtained on the basis of big data prediction, $\mathcal{V}_m^{\text{pre}} = \emptyset$, $\mathcal{V}_m{}' = \emptyset$, and the number of alternative receivers for each V2V link is set as K. Then, the vehicle v^* that makes $\delta_{m,\ v*}(t) \cdot k_{m,\ v*}(t) = 1$ is found, and v^* is put into the set $\mathcal{V}_m{}'$. Substituting $\mathcal{V}_m{}'$ into (9a) yields $\arg \max\limits_{\mathcal{V}_m{}'} R_{\text{sum}}$. When all M vehicular users finish the pairing process, the algorithm terminates.

Algorithm 3.1: Heuristic Social Mobility-Aware Vehicle Pairing Algorithm (HSMA-VPA)

Initialization: The set of social-ties $\boldsymbol{\delta}$, the transmitter sets of vehicular users and cellular users $\mathcal{M}$ and $\mathcal{U}$, the set of alternative paired vehicles $\mathcal{V}$, $\mathcal{V}_m{}' = \emptyset, \forall m \in \mathcal{M}$, the prepairing set $\mathcal{V}_m^{\text{pre}} = \emptyset, \forall m \in \mathcal{M}$, the number of alternative receivers for each V2V link K.

FOR $m = 1 : M$ **do**

Find the vehicle v^* that makes $\delta_{m,\ v*}(t) \cdot k_{m,\ v*}(t) = 1$.

Put v^* into the set $\mathcal{V}_m{}'$.

Calculate $\arg \max\limits_{\mathcal{V}_m{}'} R_{\text{sum}}$.

ENDFOR

After the vehicle pairing decision is obtained, $\mathcal{P}2$ can be converted into a spectrum assignment sub-subproblem, i.e., $\mathcal{P}4$, which is expressed as follows:

$$\begin{aligned} \mathcal{P}4: \quad & \max_{X_{s,f}(t),\, X_{s,f}(t)} \ R_{\text{sum}}, \\ \text{s.t.} \quad & (3.12b), (3.12d) - (3.12g). \end{aligned} \tag{3.15}$$

According to $\mathcal{P}4$, the relationship between a cellular user and an SC is a one-to-one match. To facilitate the solution process, the complex matches among the cellular users, vehicular users and SCs are transformed into new matches among the cellular users and vehicular users. The new matches are denoted as $X_{u,\ m}(t)$. Therefore, $\mathcal{P}4$ can be rewritten as

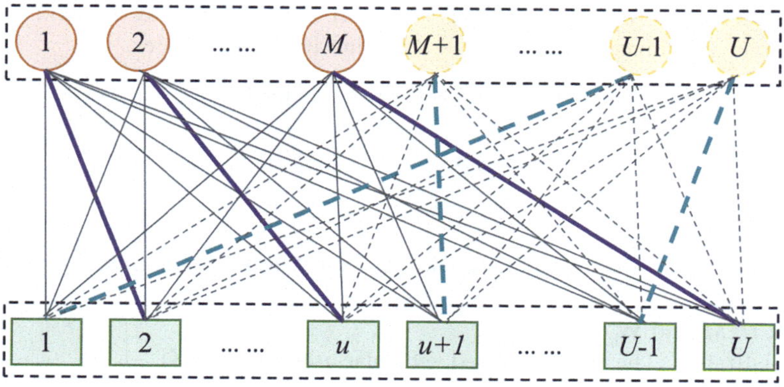

Fig. 3.2 Illustration of the revised KM-based matching method

$$\begin{aligned} \mathcal{P}5: \quad & \max_{X_{u,\alpha}(t)} R_{\text{sum}}, \\ \text{s.t.} \quad & X_{u,m}(t) \in \{0,\ 1\}, \forall\ u \in \mathcal{U}, m \in \mathcal{M}, t \in \mathcal{T}, \\ & \sum\nolimits_{m=1}^{M} X_{u,m}(t) \le 1\ ,\ \forall\ u \in \mathcal{U}, t \in \mathcal{T}, \\ & \sum\nolimits_{u=1}^{V} X_{u,m}(t) \le 1\ , \forall\ m \in \mathcal{M}, t \in \mathcal{T}. \end{aligned} \tag{3.16}$$

However, $\mathcal{P}5$ is still nonconvex. Since the complexity of exhaustive searching increases exponentially with the number of users, KM-SAA is proposed as an alternative to address the spectrum assignment subproblem, which is divided into the following two steps.

Step one involves weighted bipartite graph (WBG) construction. A weighted bipartite graph is defined as $\mathcal{G} = (\mathbf{B}, \mathbf{Y}, \mathbf{W})$, in which $\mathbf{B} = [B_1, B_2, \ldots, B_u, \ldots, B_U]$, $\mathbf{Y} = [Y_1, Y_2, \ldots, Y_m, \ldots, Y_M]$, $\mathbf{B} \cap \mathbf{Y} = \emptyset$ and $\mathbf{W} = [W_{B_u,Y_m}]_{U \times M}, B_u \in \mathbf{B}, Y_m \in \mathbf{Y}$. **B** and **Y** are the vertices of the cellular users and the vehicular users, respectively. **W** denotes the weight matrix. W_{B_u,Y_m} represents the weight between the B_uth vertex and the Y_mth vertex, which is obtained by

$$W_{B_u,Y_m} \triangleq R_{\text{sum\_sub}}, \forall B_u \in \mathbf{B}, Y_m \in \mathbf{Y}. \tag{3.17}$$

Step two is KM matching. Notably, the traditional KM algorithm requires a bipartite graph to be perfectly symmetric. However, the number of cellular users may exceed the number of vehicular users in our work. By adding $U - M$ virtual vertices with zero-weighted links, the original graph is extended to a virtual graph $\mathcal{G}'$, which is shown in Fig. 3.2. $\mathcal{G}'$ is perfectly symmetric and can be solved via KM. The key index in the KM algorithm is feasible vertex labeling, which is defined in **Definition 3.1**. The detailed process of the KM algorithm is shown in Algorithm 2. First, the two column vectors (i.e., **B** and **Y**), the weight matrix (i.e., **W**) and the vertex labeling matrix (i.e., **L**) are initialized. Then, **W**, L_{B_u} and L_{Y_m} are calculated according to (3.18). Next, if $\Pi_{\text{match}} \neq \mathcal{M}$, an augmenting path is found via the Hungarian algorithm. When $\Pi_{\text{match}} = \mathcal{M}$, Algorithm 2 terminates.

Define 3.1 A feasible vertex label in $\mathcal{G}'$ can be defined as

$$\begin{cases} L_{B_u} = \max\{W_{B_u,Y_u}\}, \forall Y_m \in \mathbf{Y}, \text{if } B_u \in \mathbf{B}, \\ L_{Y_m} = 0, \text{if } Y_m \in \mathbf{Y}. \end{cases} \tag{3.18}$$

Algorithm 3.2: Kuhn-Munkres-Based Spectrum Assignment Algorithm (KM-SAA)

Input: U, M, and $\mathbf{W}$

Initialization: The set of matched vehicular users $\Pi_{\text{match}} = \emptyset$.

Define column vectors $\mathbf{B}$ and $\mathbf{Y}$, and let $\mathbf{B} = [B_1, B_2, \ldots, B_u, \ldots, B_U]'$ and $\mathbf{Y} = [Y_1, Y_2, \ldots, Y_m, \ldots, Y_M]'$. Construct a matrix $\mathbf{W} \in \mathbb{C}^{U \times M}$. Denote the vertex labels as a matrix $\mathbf{L} = [L_{B_1}, L_{B_2}, \ldots, L_{B_u}, \ldots, L_{B_U}, L_{Y_1}, L_{Y_2}, \ldots, L_{Y_m}, \ldots, L_{Y_M}]$ and the spectrum assignment variable as $X_{u,m}(t) = 0, \ \forall u \in \mathcal{U}, m \in \mathcal{M}$.

Calculate $\mathbf{W}$, L_{B_u} and L_{Y_m}.

$L_{B_u} = \max\{W_{B_1,Y_1}, \ldots, W_{B_u,Y_m}, \ldots, W_{B_U,Y_M}\}, \ \forall B_u \in \mathbf{B}$.

WHILE $\Pi_{\text{match}} \neq \mathcal{M}$

Find an augmenting path using the Hungarian algorithm, and then update Π_{match}.

ENDWHILE

Remove the virtual vertices and edges in $\mathcal{G}'$.

Output: The spectrum assignment decision $X_{u,m}(t)$.

3.3.2 3D Power Control Algorithm

After the vehicle pairing and spectrum assignment strategies are obtained, the power control subproblem can be expressed as

$$\mathcal{P}6: \max_{P_{m,1}(t), P_{m,2}(t), P_u(t)} R_{\text{sum}}, \tag{3.19a}$$

$$\text{s.t.} B_{sc} \log_2 \left(1 + \frac{P_{m,2}(t) X_{m,f}(t) H_{m,v,2}(t)}{\sum_{u \in \mathcal{U}} P_u(t) X_{u,f}(t) H_{u,m,2}(t) + \sigma^2}\right) \geq R^m_{th,2}, \forall m \in \mathcal{M}, \forall v \in \mathcal{V}^{\text{pre}}_m, f \in \mathcal{F}, t \in \mathcal{T}, \tag{3.19b}$$

$$\frac{\sum_{v \in \mathcal{V}^{\text{pre}}_m} P_{m,1}(t) H_{m,v,2}(t) - P_{m,2}(t) H_{m,v,2}(t)}{\sigma^2} \geq \xi, \forall m \in \mathcal{M}, t \in \mathcal{T}, \tag{3.19c}$$

$$(3.12h), (3.12j).(3.12l), (3.12m). \tag{3.19d}$$

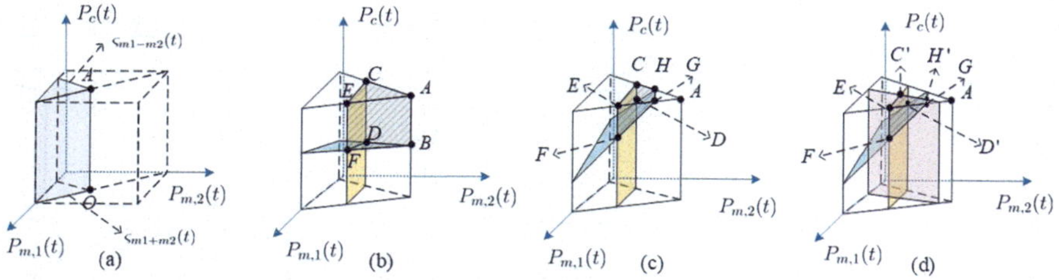

Fig. 3.3 (**a**) The shaded portions represent the condition in which constraints (3.12l), (3.12m), and (3.19b) are satisfied. (**b**)-(**d**) Three feasible power domain cases are observed. (**b**) Point *A* satisfies all the constraints. (**c**) Point *A* does not satisfy constraint (3.12h). (**d**) Point *A* does not satisfy constraints (3.12h) and (3.12j)

The feasible region of $\mathcal{P}6$ should be the intersection of all the constraints. To facilitate the solution process, the feasible region is expressed in three-dimensional coordinates, as shown in Fig. 3.3a, where the axis coordinates are $P_{m,\,1}$, $P_{m,\,2}$ and P_u. When (3.12m) and (3.19c) are satisfied (i.e., the left-hand side of the constraints is equal to the right-hand side), inequalities (3.12m) and (3.19c) can be transformed into

$$\varsigma_{m1+m2}(t) \triangleq P_{m,1}(t) + P_{m,2}(t) - P_m^{\max} = 0\ , \tag{3.20}$$

$$\varsigma_{m1\,-m2}(t) \triangleq \sum_{v\in \mathcal{V}_m^{\text{pre}}} \left(P_{m,1}(t)H_{m,v,2}(t) - P_{m,2}(t)H_{m,v,2}(t)\right) - \varrho\sigma^2 = 0. \tag{3.21}$$

As shown in Fig. 3.3a, line AO is the intersection of planes $\varsigma_{ml\ -\ m2(t)}$ and $\varsigma_{m1\ +\ m2}(t)$. Therefore, the intersection of (3.12l), (3.12m), and (3.19c) is shown in the shaded portion of Fig. 3.3a. When the equalities of (3.12h), (3.12j), and (3.19b) are satisfied, inequalities (3.12h), (3.12j), and (3.19b) can be substituted by

$$\varsigma_{m,1}(t) \triangleq -P_{m,1}(t)H_{m,s}(t) - P_u(t)H_{u,m}(t)\gamma_{th,1}^m - P_{m,2}(t)H_{m,s}(t)\gamma_{th,1}^m - \sigma^2\gamma_{th,1}^m = 0\ , \tag{3.22}$$

$$\varsigma_u(t) \triangleq \sim P_u(t)H_u(t) - P_{m,1}(t)H_{m,u}(t)\gamma_{th}^u - P_{m,2}(t)H_{m,v,2}(t)\gamma_{th}^u - \sigma^2\gamma_{th}^u = 0\ , \tag{3.23}$$

$$\varsigma_{m,2}(t) \triangleq \sim P_{m,2}(t)H_{m,2}(t) - P_u(t)H_{u,v,2}(t)\gamma_{th,2}^m - \sigma^2\gamma_{th,2}^m = 0\ , \tag{3.24}$$

where $\gamma_{th,1}^m = 2^{\frac{R_{th,1}^m}{B_{sc}}} - 1$, $\gamma_{th}^u = 2^{\frac{R_{th}^u}{B_{sc}}} - 1$ and $\gamma_{th,2}^m = 2^{\frac{R_{th,2}^m}{B_{sc}}} - 1$. Figure 3.3 shows that Eqs. (3.22)–(3.24) are represented as blue, pink, and yellow planes, respectively. Notably, the slope and size of each plane are jointly decided by the transmission

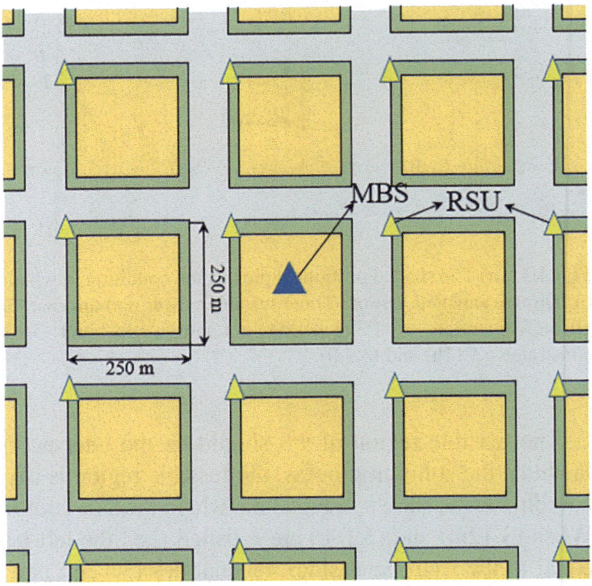

Fig. 3.4 Simulation setups for the road configuration in the urban scenario

powers and QoS constraints, as well as the channel gains. Therefore, the intersection of (3.19b)–(3.19d) is shown in the shaded portions of Fig. 3.3b–d.

Through the above analysis, the feasible region of the transmission power varies with the slopes of the three planes. Three cases are observed for the feasible power domain. Case 1: When the slopes of the three planes are small, point $A\ (P_{m,1}(t), P_{m,2}(t), P_u(t)) = ([\varsigma_{m1-m2}(t), \varsigma_{m1+m2}(t)]|_{[P_{m,1}(t),P_{m,2}(t)]}, P_u^{\max})$ guarantees all the constraints. The power region is illustrated in Fig. 3.3b.

Case 2: Together with the increase in the slope of a plane, point A may no longer satisfy one of the constraints among (3.12h), (3.12j), or (3.19b). Notably, $P_u(t)$ is not included in constraint (3.19b). Therefore, constraint (3.19b) is always satisfied at point A, and there are only two cases in which constraint (3.12h) or (3.12j) cannot be guaranteed at point A (as shown in Fig. 3.3c). Case 3: As the slopes of any two planes continually increase, point A will not guarantee two out of the three constraints (3.12h), (3.12j), and (3.19b). Thus, there is only one case where constraints (3.12h) and (3.12j) cannot be satisfied at point A (as shown in Fig. 3.3d). As proven in [8], the optimal transmission power $\left(P_{m,1}^*(t), P_{m,2}^*(t), P_u^*(t)\right)$ of $\mathcal{P}3$ is found at the vertices of the feasible region.

Therefore, for Case 1, the alternative transmission power solutions at the vertices can be expressed as

$$
\begin{aligned}
A&: ([P_{m,1}(t), P_{m,2}(t)]|_{[\varsigma_{m1-m2}(t),\varsigma_{m1+m2}(t)]}, P_u^{\max}),\\
B&: ([P_{m,1}(t), P_{m,2}(t), P_u(t)]|_{[\varsigma_{m1-m2}(t),\varsigma_{m1+m2}(t),\varsigma_{m,1}(t)]}),\\
C&: ([P_{m,1}(t), P_{m,2}(t)]|_{[\varsigma_{m,2}(t),\varsigma_{m1-m2}(t)]}, P_u^{\max}),\\
D&: ([(P_{m,1}(t), P_{m,2}(t), P_u(t)]|_{[\varsigma_{m1-m2}(t),\varsigma_{m,1}(t),\varsigma_{m,2}(t)]}),\\
E&: ([P_{m,1}(t), P_{m,2}(t)]|_{[\varsigma_{m,2}(t),\varsigma_{m1+m2}(t)]}, P_u^{\max}),\\
F&: ([P_{m,1}(t), P_{m,2}(t), P_u(t)]|_{[\varsigma_{m,2}(t),\varsigma_{m,1}(t),\varsigma_{m1+m2}(t)]}).
\end{aligned}
$$

In Case 2, if point A does not guarantee constraint (3.12h), as shown in Fig. 3.4c, the alternative transmission powers at the vertices can be denoted as

$$
\begin{aligned}
C&: ([P_{m,1}(t), P_{m,2}(t)]|_{[\varsigma_{m,2}(t),\varsigma_{m1-m2}(t)]}, P_u^{\max}),\\
D&: ([P_{m,1}(t), P_{m,2}(t), P_u(t)]|_{[\varsigma_{m1-m2}(t),\varsigma_{m,1}(t),\varsigma_{m,2}(t)]}),\\
E&: ([P_{m,1}(t), P_{m,2}(t)]|_{[\varsigma_{m,2}(t),\varsigma_{m1+m2}(t)]}, P_u^{\max}),\\
F&: ([P_{m,1}(t), P_{m,2}(t), P_u(t)]|_{[\varsigma_{m,2}(t),\varsigma_{m,1}(t),\varsigma_{m1+m2}(t)]}),\\
G&: ([P_{m,1}(t), P_{m,2}(t)]|_{[\varsigma_{m,1}(t),\varsigma_{m1+m2}(t)]}, P_u^{\max}),\\
H&: ([P_{m,1}(t), P_{m,2}(t)]|_{[\varsigma_{m,1}(t),\varsigma_{m1-m2}(t)]}, P_u^{\max})).
\end{aligned}
$$

Similarly, if point A does not guarantee constraint C9, an alternative power can be obtained. Case 3 illustrates that the alternative transmission power solutions at the vertices can be given by

$$
\begin{aligned}
C'&: ([P_{m,1}(t), P_{m,2}(t)]|_{[\varsigma_u(t),\varsigma_{m,2}(t)]}, P_u^{\max}),\\
D'&: ([P_{m,1}(t), P_{m,2}(t), P_u(t)]|_{[\varsigma_u(t),\varsigma_{m,2}(t),\varsigma_{m,1}(t)]}),\\
E&: ([P_{m,1}(t), P_{m,2}(t)]|_{[\varsigma_{m,2}(t),\varsigma_{m1+m2}(t)]}, P_u^{\max}),\\
F&: ([P_{m,1}(t), P_{m,2}(t), P_u(t)]|_{[\varsigma_{n,2}(t),\varsigma_{m,1}(t),\varsigma_{m1+m2}(t)]}),\\
G&: ([P_{m,1}(t), P_{m,2}(t)]|_{[\varsigma_{m,1}(t),\varsigma_{m,1+m2}(t)]}, P_u^{\max}\}),\\
H'&: ([P_{m,1}(t), P_{m,2}(t)]|_{[\varsigma_s(t),\varsigma_{m,1}(t)]}, P_u^{\max}).
\end{aligned}
$$

By substituting the above solutions into $\mathcal{P}3$, $P_{m,1}^*(t)$, $P_{m,2}^*(t)$, and $P_u^*(t)$ can be solved via limited searches.

Algorithm 3.3: NOMA-Based Joint Vehicle Pairing, Spectrum Assignment and Power Control Algorithm (NOMA-JVP-SA-PCA)

Input: $\mathcal{U}$: set of cellular users, $\mathcal{M}$: set of vehicular users.

WHILE no convergence

With the given power control strategy, perform Algorithm 3.1 and Algorithm 3.2 to calculate the sum rate.

With the given vehicle pairing and spectrum assignment decision, perform 3D-PCA to calculate the sum rate.

ENDWHILE

Output: Vehicle pairing decision $\Psi_{m,v}(t)$, spectrum assignment strategy $\mathbf{X}$ and transmission power $\mathbf{P}$.

3.3.3 *Description of the Proposed Algorithm*

In conclusion, to solve $\mathcal{P}_1$, the original sum-rate optimization problem is decoupled into two subproblems, i.e., the joint vehicle pairing and spectrum assignment subproblem and the power control subproblem. Algorithm 3.1 and Algorithm 3.2 are proposed to solve the first subproblem, and then, with the given vehicle pairing and spectrum assignment strategy, a low-complexity 3D-PCA approach is proposed. Finally, through an iteration process, the vehicle pairing variables, spectrum assignment variables, and transmission power variables are iteratively updated until convergence is achieved. The specific procedure is shown in Algorithm 3.3.

3.3.4 *Complexity Analysis of the Proposed Algorithm*

The proposed NOMA-JVP-SA-PCA consists of two parts, joint vehicle pairing and spectrum assignment design and power control decision-making, which are iteratively executed. The number of iterations is defined as N_1. In the joint vehicle pairing and spectrum assignment stage, Algorithm 3.1 and Algorithm 3.2 are iteratively executed. The complexity of Algorithm 1 is $\mathcal{O}(M)$, and the complexity of Algorithm 3.2 can be approximated as $\mathcal{O}(\chi^3)$ [9], where χ is $\max\{U, M\}$. Since $M \ll \chi^3$, the complexity at this stage is approximately $\mathcal{O}(N_2\chi^3)$, where the number of iterations is N_2. In the power control stage, for each SC, the complexities of Case 1, Case 2 and Case 3 are approximately $\mathcal{O}(6K)$, $\mathcal{O}(12K)$ and $\mathcal{O}(6K)$, respectively. Therefore, the total complexity of the proposed power control strategy for all F SCs is $\mathcal{O}(24KF)$. Thus, the total complexity of the proposed NOMA-JVP-SA-PCA is $\mathcal{O}(N_1(24KF + N_2\chi^3))$. In comparison, the complexity of exhaustive searching is $(2 + V)^M N_p^{(U+M)}$, where N_p is the number of power levels that each transmitter can choose from. The complexity of the proposed NOMA-JVP-SA-PCA is much lower than that of the exhaustive search method.

3.4 Simulations and Discussions

When performing the simulations, we consider a rectangular area with an MBS located at the center and numerous RSUs located at the intersections of the streets within the area, as shown in Fig. 3.4. The parameters are set according to 3GPP TR 36.885. We assume that all the RSUs are located close to the road and that the distance between two adjacent RSUs is 200 m. The average intervehicle distance in the same lane is 2.5v m, and v denotes the moving speed of a vehicle in meters per second [11]. U cellular users are randomly located within the area. To perform authentic simulations, the public database of the Stanford Network Analysis Platform (SNAP) is adopted to model social tiers. Moreover, the channel parameters are set according to [6, 7, 12, 13]. The remaining parameters are listed in Table 3.1.

3.4.1 Superiority Analysis of the Proposed Joint Social-Physical Domain Vehicle Pairing Scheme

To reveal the necessity of the proposed vehicle pairing scheme for achieving a high sum rate, the minimum distance-based vehicle pairing (D-VP) strategy in [14] is adopted as a baseline. To conduct fair comparisons, the baseline is carried out by integrating the proposed spectrum assignment algorithm and the power control strategy with NOMA. The simulations are performed by setting $P_m^{\max} = 17$ dBM,

Table 3.1 Simulation parameters and values

Parameters	Values
Maximum cellular transmission power $P_u^{\max}$	30 dBm
Maximum vehicle transmission power $P_m^{\max}$	[30, 50] dBm
System bandwidth	15 MHz
Carrier center frequency	2 GHz
Vehicle speed	[20, 90] km/h
Number of cellular users U	[10, 50]
Number of vehicular users M	[10, 50]
Maximum transmission range of V2V links	50 m
Number of lanes	6
Lane width	4
SINR threshold of cellular links γ_{th}^{u}	5 dB
SINR threshold of V2V links $\gamma_{th,2}^{m}$	0 dB
SINR threshold of V2I links $\gamma_{th,1}^{m}$	[0, 5] dB
Shadowing standard deviation for cellular links	8 dB
Shadowing standard deviation for V2I and V2V links	3 dB
Required power difference ratio ξ	[6, 10] dB

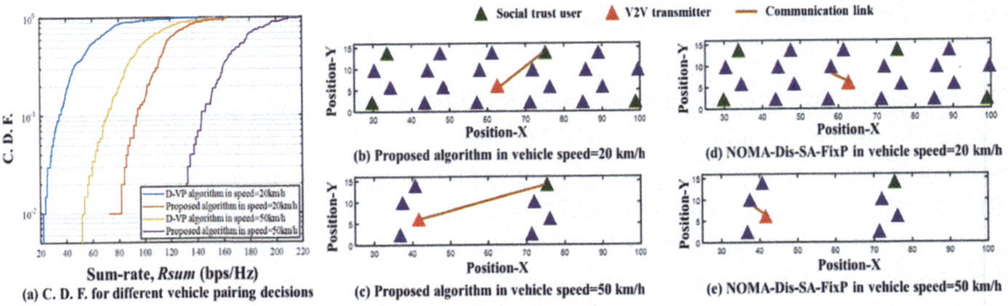

Fig. 3.5 System sum rates of different vehicle pairing policies

$\gamma_{th,1}^{m} = 0$ dB, $\xi = 6$ dB and $U = M = 30$. As shown in Fig. 3.5a, the proposed vehicle pairing scheme outperforms D-VP. This is because the proposed algorithm can precisely avoid building communication links among vehicle pairs with no social ties, as shown in Fig. 3.5b–e. In this way, the degree of spectrum resource waste is reduced. In contrast, D-VP generates many communication links among vehicle users with no social ties, resulting in high spectrum resource occupation and finally decreasing the sum rate. This outcome convincingly confirms that the proposed vehicle pairing scheme is more effective than D-VP in improving the sum rate.

3.4.2 *Sum-Rate Performance*

In this subsection, we study the impacts of the parameters on the sum-rate performance of the utilized system. To demonstrate the superiority of the proposed NOMA-JVP-SA-PCA, three algorithms are simulated and compared.

1. NOMA-Dis-SA-FixP algorithm. This algorithm is composed of the minimum distance-based vehicle pairing strategy and the auction-based spectrum assignment algorithm with a fixed maximum transmission power from [10].
2. NOMA-JVP-RA-PC algorithm. This algorithm is composed of the random spectrum assignment method and the proposed JVA-PCA.
3. NOMA-JVP-RA-FixP algorithm. This algorithm is composed of the proposed JVA-PCA and the random spectrum assignment method with a fixed maximum transmission power.

The impact of the number of available SCs on the sum rate is simulated by setting $P_m^{\max} = 40$ dBm, $\gamma_{th,1}^{m} = 0$ dB and $\xi = 6$ dB, and the obtained results are shown in Fig. 3.6. The proposed NOMA-JVP-SA-PCA achieves a higher sum rate than the baselines do, and the gains are at least 20%. This also confirms that the NOMA-JVP-SA-PCA can effectively enhance the sum-rate performance of V2X systems. In addition, Fig. 3.7 shows that the sum rate increases linearly with the number of available SCs. This is because the number of accessible vehicles increases linearly as the number of available SCs increases, which ultimately enhances the system sum rate.

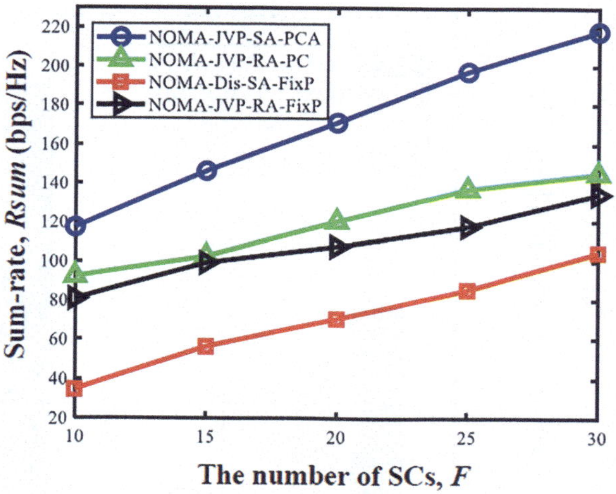

Fig. 3.6 The sum rates obtained with different numbers of SCs

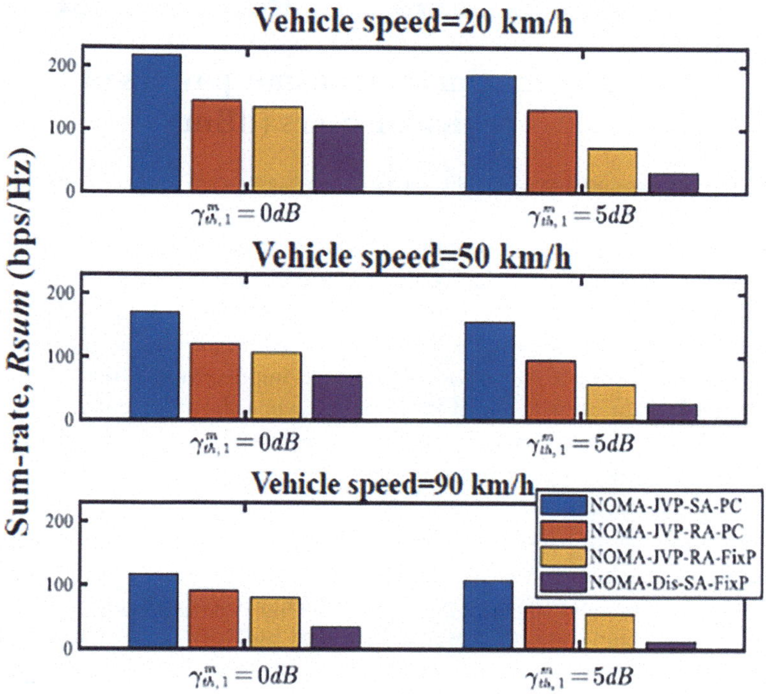

Fig. 3.7 The sum rates obtained with different vehicle speeds

The impact of the vehicle speed on the sum rate is investigated under different SINR thresholds for V2I links, i.e., $\gamma_{th,1}^{m}$. The simulations are performed by setting $P_m^{\max} = 40$ dBm and $\xi = 6$ dB. The simulation results are presented in Fig. 3.7, which shows that the sum rate decreases when the vehicle speed increases from

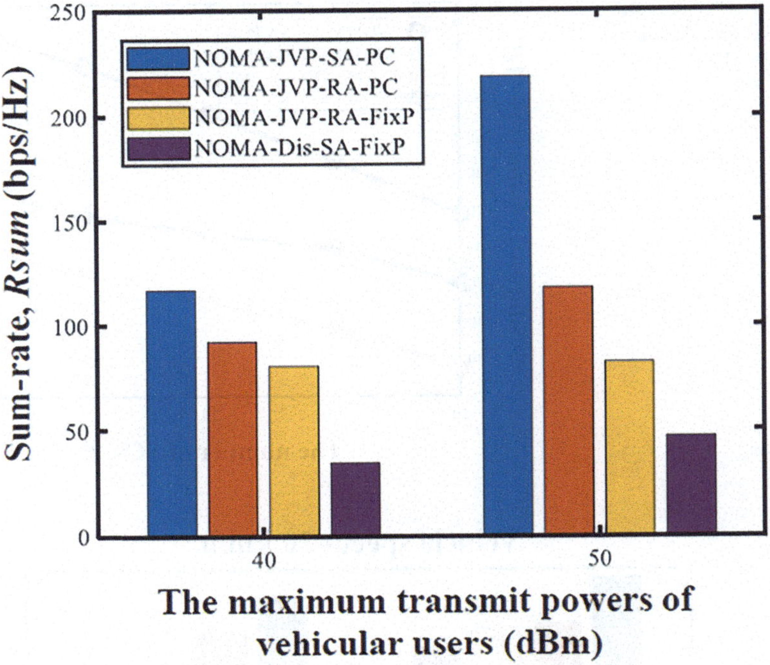

Fig. 3.8 The sum rates obtained with different maximum transmission powers for vehicular users

20 to 90 km/h for all the algorithms. This is because when the vehicle speed increases, the number of vehicles contained in the fixed-length streets decreases. In addition, the performance gap between the proposed NOMA-JVP-SA-PCA and the baselines gradually decreases as the vehicle speed increases. This also illustrates that the proposed algorithm provides a greater improvement over the baseline algorithms when the vehicle speed is slow, i.e., when the vehicle density is high. Moreover, the sum rate decreases with increasing $\gamma_{th,1}^{m}$. This is because with increasing $\gamma_{th,1}^{m}$, it becomes difficult to satisfy the constraints imposed by differentiated QoS requirements.

The sum-rate performance for different maximum transmit powers of vehicular users is also investigated. The results are included in Fig. 3.8, which shows that enhancing the maximum transmit power of vehicular users can improve the sum-rate. When maximum transmit power is increased by 1 dBm, the sum-rate performance can be enhanced about 10 bps/Hz by the proposed NOMA-JVP-SA-PCA. Although enhancing the transmit power may also cause increasing co-channel interference, it can be ignored compared with the power enhancement of the desired signal. To compare, we also simulate the fixed power transmitted case. It is observed that with the increase of the maximum transmit powers, the sum rate of baselines keeps almost the same.

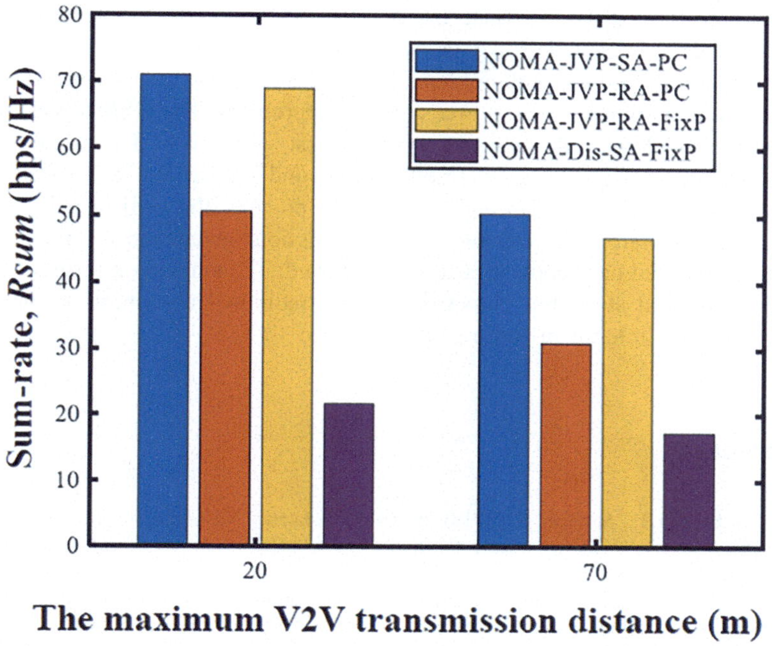

Fig. 3.9 The sum rates obtained with different maximum V2V transmission distances

Finally, the effect of the V2V transmission distance on the sum-rate performance is investigated by setting $P_m^{\max} = 40$ dBm, $\gamma_{th,1}^m = 0$ dB, $\xi = 6$ dB and $U = M = 5$. The simulation results are shown in Fig. 3.9. As the V2V transmission distance increases, the sum-rate curves of both the proposed algorithm and the baselines dramatically decrease. This is because as the V2V transmission distance increases, the channel quality decreases, which reduces the sum-rate performance of the vehicular networks. In addition, the proposed NOMA-JVP-SA-PCA has the highest sum-rate performance. Specifically, when the maximum V2V transmission distance is 70 m, the system sum-rate performance gain provided by the proposed algorithm is at least 10%. This also confirms that the proposed NOMA-JVP-SA-PCA can effectively enhance the sum-rate performance of V2X systems.

3.5 Summary

This chapter focused on social mobility-aware sum-rate maximization for NOMA-enabled underlaid C-V2X networks. Taking the diverse QoS requirements of both cellular users and vehicular users into account, we formulated a sum-rate maximization problem by jointly optimizing the vehicle pairing, spectrum assignment and power control situations. Because the problem was observed to be NP-hard, an iterative method, NOMA-JVP-SA-PCA, was proposed to solve it by decoupling the original problem into two subproblems, i.e., a joint vehicle pairing and spectrum assignment subproblem and a power control subproblem. The HSMA-VPA and a revised KM-SAA were designed to solve the first subproblem. Given the vehicle pairing and spectrum assignment results, the second subproblem was solved via a low-complexity 3D-PCA. Through an iterative process, the vehicle pairing variables, spectrum assignment variables, and transmission power control variables were iteratively updated until convergence was achieved. The simulation results showed that the proposed algorithm outperformed the traditional algorithms by approximately 24–53% in terms of sum-rate performance.

References

1. Abbas, F., Fan, P., & Khan, Z. (2019). A novel low-latency V2V resource allocation scheme based on cellular V2X communications. *IEEE Transactions on Intelligent Transportation Systems, 20*(6), 2185–2197.
2. Hu, B., & Chu, X. (2021). Social-aware resource allocation for vehicle-to-everything communications underlaying cellular networks. In *2021 IEEE 93rd Vehicular Technology Conference (VTC2021-Spring), Helsinki, Finland* (pp. 1–6).
3. Zhang, Y., & Zhou, Y. (2023). Resource allocation strategy based on tripartite graph in vehicular social networks. *IEEE Transactions on Network Science and Engineering, 10*(5), 3017–3031.
4. Wu, D., Li, J., He, P., Cui, Y., & Wang, R. (2023). Social-aware graph-based collaborative caching in edge-user networks. *IEEE Transactions on Vehicular Technology, 72*(6), 7926–7941.
5. Guo, S., & Zhou, X. (2019). Robust resource allocation with imperfect channel estimation in NOMA-based heterogeneous vehicular networks. *IEEE Transactions on Communications, 67*(3), 2321–2332.
6. Chen, C., Wang, B., & Zhang, R. (2019). Interference hypergraph-based resource allocation (IHG-RA) for NOMA-integrated V2X networks. *IEEE Internet of Things Journal, 6*(1), 161–170.
7. Zhang, F., Wang, M. M., Bao, X., & Liu, W. (2021). Centralized resource allocation and distributed power control for NOMA-integrated NR V2X. *IEEE Internet of Things Journal, 8*(22), 16522–16534.
8. Li, X., Zhang, J., & Liu, X. (2023). Mode selection and resource allocation for D2D communications with tradeoff of throughput and user satisfaction. *IEEE Transactions on Communications, 71*(9), 5607–5620.
9. Peng, X., Ota, K., & Dong, M. (2020). Multi-attribute-based double auction toward resource allocation in vehicular fog computing. *IEEE Internet of Things Journal, 7*(4), 3094–3103.
10. Wu, W., Xue, T., Wang, Q., Han, S., & Wu, X. (2019). A two-stage resource allocation for scma-based c-v2x networks. In *2019 IEEE Global Communications Conference (GLOBECOM)* (pp. 1–6).
11. Zhao, B., Liu, J., Mao, B., & Li, S. (2023). Optimal resource allocation for random multiple access oriented SCMA-V2X networks. *IEEE Transactions on Vehicular Technology, 72*(8), 10921–10932.

12. Sun, W., Yuan, D., Ström, E. G., & Brännström, F. (2016). Cluster-based radio resource management for D2D-supported safety-critical V2X communications. *IEEE Transactions on Wireless Communications, 15*(4), 2756–2769.
13. Ding, H., & Leung, K.-C. (2021). Resource allocation for low-latency NOMA-V2X networks using reinforcement learning. In *IEEE INFOCOM 2021 – IEEE Conference on Computer Communications Workshops (INFOCOM WKSHPS), Vancouver, BC, Canada* (pp. 1–6).
14. Chai, G., Wu, W., Yang, Q., Qin, M., Wu, Y., & Yu, F. R. (2024). Platoon partition and resource allocation for ultra-reliable V2X networks. *IEEE Transactions on Vehicular Technology, 73*(1), 147–161.

Chapter 4
Socially Aware Caching Resource Management in VSNs

Newly emerging services, such as automatic driving and intelligent transportation, have strict requirements for network latency in vehicular networks [1, 2]. To address this, it is necessary to introduce advanced networking architectures and efficient data transmission techniques for next-generation mobile networks and beyond (i.e., B5G/6G). Edge caching has been regarded as a promising technique to relieve such network transmission pressures and reduce the resulting network latency [3]. By precaching popular content near the edges of networks, vehicles can access their required content directly from much closer roadside units (RSUs) or adjacent vehicles [4–7]. This can remarkably decrease the content delivery latency and greatly improve the experience. However, in real life, openly accepting the content sent by others may be very dangerous since this content may contain viruses. For security reasons, vehicles usually accept only content sent by familiar sources [8–10]. Inspired by this, the social relationships among vehicles (i.e., vehicular social networks–VSNs) need to be investigated to help choose important vehicles (IVs) as caching service nodes. In addition, as the caching storage of vehicles is normally limited, the content that should be cached becomes the main issue. At present, the main way to decide what to cache heavily relies on the popularity of the content, which is typically unknown in advance owing to the high dynamics of content requests. However, the prediction methods proposed in existing works are based on centralized learning, which requires vehicles to send content request information or even more personal information (such as age and gender) to a central server to train the model [11–13]. This may cause serious privacy and security issues, so vehicles may be unwilling to share such information with the central server, which in turn hinders the application of proactive edge caching. Therefore, there is an urgent need to design a distributed learning framework to predict content popularity, where data with personal information are stored and used only at the user side. Against this background, socially aware caching resource management in VSNs is proposed in this chapter. The main contributions of this study are summarized as follows:

H. Zhang et al., *Intelligent Resources Management for Vehicular Social Networks*, Wireless Networks, https://doi.org/10.1007/978-3-031-80169-3_4

- Edge caching in VSNs is investigated by considering dynamic content popularity, vehicle mobility, and social properties. We formulate an optimization problem to maximize the system benefit per unit cost. Since the problem is NP-hard and difficult to solve, a novel community detection and attention-weighted federated learning-based proactive edge caching (CAFLPC) strategy is proposed to obtain the near-optimal solution by decomposing the whole edge caching process into two stages, i.e., caching node selection and caching content placement.
- To select the optimal IVs as caching nodes, a novel vehicle-to-vehicle (V2V) sharing graph-based community detection method is proposed. To our knowledge, this is the first work that jointly considers vehicle mobility and social properties when users are grouped into different communities.
- To place the popular content on the selected IVs without privacy disclosure, an attention-weighted federated learning (AWFL)-based content popularity prediction method is proposed. With this method, an AWFL framework for content popularity prediction is proposed by adopting bidirectional long short-term memory (BiLSTM) as a prediction model. Most importantly, considering the imbalance problem of vehicle model quality, an attention-weighted aggregation mechanism is proposed by jointly considering the activity levels and local training epochs of vehicles to generate different aggregation weights for different local quality models in the FL aggregation step.
- The simulation results on a real-world dataset show that, compared with the typical FL framework, the proposed AWFL framework can improve the accuracy of content popularity prediction by approximately 0.68–4.19% in terms of the mean absolute error. Moreover, the proposed CAFLPC strategy outperforms the simulated existing caching strategies by approximately 2.2–35.1% in terms of the transmission latency, which is reduced per unit cost.

4.1 Framework of Cache-Enabled VSNs

4.1.1 Network Architecture

As shown in Fig. 4.1, a cache-enabled VSN consisting of one RSU and U randomly distributed vehicles is studied. In the system, each vehicle in set $\mathcal{U} = \{1, 2, \ldots, u, \ldots, U\}$ can request any content from the content library $\mathcal{F} = \{1, 2, \ldots, f \ldots, F\}$, in which the length of content f is denoted as l_f. To avoid possible traffic congestion caused by repetitive content downloading requests, some vehicles are hired as IVs to provide caching services. By precaching those popular contents, the selected IVs can transmit the desired contents to the requesting vehicles via V2V communications. The set of these hired IVs is denoted by $I = \{1, 2, \ldots, i, \ldots I\}$, where i represents the index of the ith IV. The caching capacity of IV i is denoted by S_i. The caching states of all the IVs are denoted by $X = \left(x_{i,f}^t\right)_{I \times F}$, where $x_{i,f}^t = 1$ if content f is cached in IV i at time slot t; otherwise, $x_{i,f}^t = 0$. To simplify the model, we assume that one vehicle can be served by only

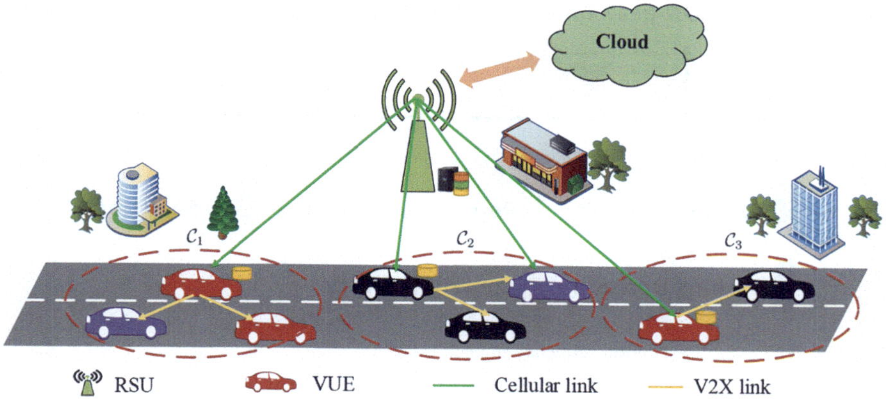

Fig. 4.1 Illustration of the cache-enabled VSN architecture

one IV in the system. The set of vehicles served by IV i is denoted as $\mathcal{C}_i$. Moreover, because the RSU can easily retrieve any content from the core networks via cables, we assume that the caching memory of the RSU is large enough to store the entire content library $\mathcal{F}$ [14].

When vehicle u sends a request to access content f, the desired content f can be transmitted from the servicing IU i or BS. Assuming that there is a resource scheduler that allocates subchannels to IVs dynamically to avoid interference, the transmission rates of the content f from the IU i and RSU to the vehicle u can be expressed as:

$$R_{i,u} = B_{\text{V2V}} \log_2\left(1 + \frac{P_{\text{VE}}|h_{i,u}|^2 d_{i,u}^{-\alpha}}{\sigma^2}\right), \tag{4.1}$$

$$R_{s,u} = B_{\text{RSU}} \log_2\left(1 + \frac{P_{\text{RSU}}|h_{s,u}|^2 d_{s,u}^{-\alpha}}{\sigma^2}\right), \tag{4.2}$$

where B_{V2V} and B_{RSU} denote the bandwidths allocated to the channels for V2V transmission and cellular links, respectively; P_{VE} and P_{RSU} denote the transmit powers of the IV and RSU, respectively; $h_{i,\ u}$ and $h_{s,\ u}$ represent the Rayleigh fast fading coefficients of the communication link for the IV i to the vehicle u and the RSU to the vehicle u, respectively; $d_{i,\ u}$ and $d_{s,\ u}$ denote the physical distances from the vehicle u to the IV i and RSU, respectively; α is the path-loss exponent; and σ^2 represents the noise power. With these transmission rates, the latency of vehicle u downloading the content f from the RSU and the IV i can be written as:

$$\tau_{s,u}^{f} = l_f / R_{s,u}, \tag{4.3}$$

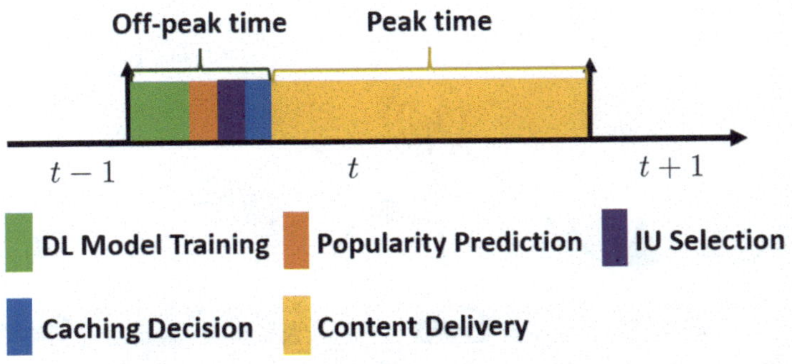

Fig. 4.2 The overall execution sequence in each time slot

$$\tau_{i,u}^{f} = l_f / R_{i,u}. \tag{4.4}$$

Considering that the distance between two vehicles communicating via V2V is typically shorter than that between an RSU and a vehicle, we assume that the content downloading latency $\tau_{i,u}^{f} < \tau_{s,u}^{f}$ when vehicle u can establish a direct V2V link with IV i.

The overall execution sequence of the proposed method in each time slot is shown in Fig. 4.2. Each time slot is divided into two phases: the off-peak phase and the peak phase. At the beginning of each time slot, the deep learning (DL) model is trained on the historical content request data. Then, IVs are selected via the proposed method, and the popularity of the contents is predicted by the trained model. On the basis of IV selection and content popularity prediction, we obtain the caching decision and update the cached contents on all IVs accordingly. During peak hours, the contents are delivered according to vehicle requests. To accurately predict the dynamic content popularity, we adopt the sliding approach to train the proposed DL model, i.e., after one time slot, the vehicle's content request data during peak hours are utilized to train the model during the off-peak hours of the next time slot.

4.1.2 V2V Sharing Model

To enable the aforementioned V2V transmission process, the V2V link should be established first. Owing to physical transmission power constraints and social trust considerations, a V2V communication link is considered successfully established only if the distance between the two vehicles is within the maximal V2V communication range and if the involved pair of vehicles has a direct social relationship. In this way, the V2V sharing pattern between two vehicles is modeled on the basis of both the physical and social domains, as shown in Fig. 4.3. The details of these two domains are discussed below.

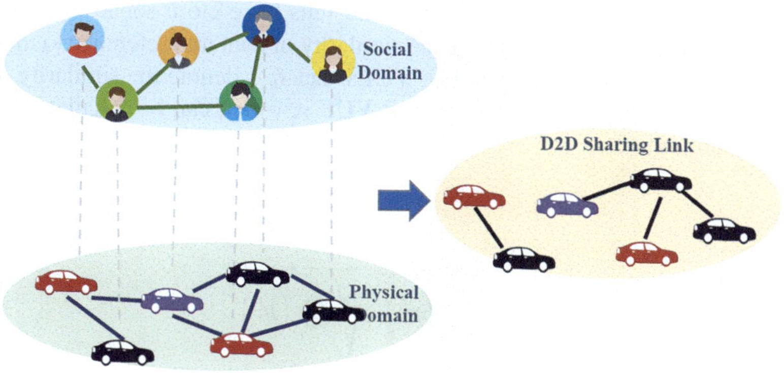

Fig. 4.3 Illustration of the V2V sharing model based on physical and social domains

1. *Physical Domain*: The physical distance between two VUEs always changes due to their mobile nature. To characterize the user's mobility pattern, the intercontact model is adopted [15]. In particular, the timeline of a VUE contact process alternatively consists of the contact time and the intercontact time. The former is the time when two VUEs are in the communication range and may share the required content via V2V links. The latter is the time between two consecutive contacts. By analyzing the actual user traces, the contact time and intercontact time between VUEs u and v follow an exponential distribution with parameters $\lambda_{u,\ v}$ and $\mu_{u,\ v}$, respectively. We define the binary random variable $m_{u,v}^{t}$ as a connection indicator for VUE u and VUE v at time slot t. During the time slot t, if VUE u and VUE v can establish a V2V communication link to transmit the requested contents, $m_{u,v}^{t}=1$; otherwise, $m_{u,v}^{t}=0$. According to [16], the connection between VUE u and VUE v follows the continuous-time Markov chain (CTMC) model, which can be expressed as:

$$\mathbb{P}\left(m_{u,v}^{t}=1|m_{u,v}^{t_0}\right)=\begin{cases}\dfrac{\mu_{u,v}-\mu_{u,v}e^{-\left(\lambda_{u,v}+\mu_{u,v}\right)(t-t_0)}}{\lambda_{u,v}+\mu_{u,v}}, if\ \ m_{u,v}^{t_0}=0,\\ \text{[2mm]}\dfrac{\mu_{u,v}+\lambda_{u,v}e^{-\left(\lambda_{u,v}+\mu_{u,v}\right)(t-t_0)}}{\lambda_{u,v}+\mu_{u,v}}, if\ \ m_{u,v}^{t_0}=1,\end{cases} \tag{4.5}$$

   where t_0 is the initial time slot.
2. *Social Domain*: Intuitively, owing to the selfish nature of human beings, VUEs are more willing to share their own content with those with whom they have stronger social relationships. This phenomenon is called social trust. On this basis, a social graph $\mathcal{G}_S=\{\mathcal{U},\mathcal{E}_S\}$ is adopted, where $\mathcal{U}$ represents the set of VUEs and where $\mathcal{E}_S=\{e_{u,v},\forall u,v\in\mathcal{U}\}$ is the set of social relationships among VUEs. When two VUEs have close social relationships, $e_{u,\ v}=1$; otherwise, $e_{u,\ v}=0$. In general, VUEs with common content preferences have a greater

probability of joining common groups and thus have close social relationships [17]. Thus, in this work, we utilize the similarity of content preference to measure the social relationships between two VUEs. Let $s_{u,\ v}$ denote the similarity of the content preference between VUE u and VUE v, which can be calculated by the cosine similarity as:

$$s_{u,v} = \frac{\sum_{f=1}^{F} p_u^f p_v^f}{\sqrt{\sum_{f=1}^{F} \left(p_u^f\right)^2} \cdot \sqrt{\sum_{f=1}^{F} \left(p_v^f\right)^2}}, \tag{4.6}$$

where p_u^f and p_v^f represent the preferences of VUE u and v for the content f. If the similarity of content preference between VUE u and VUE v is greater than a certain threshold θ, VUE u and VUE v have a close social relationship, i.e., when $s_{u,\ v} \geq \theta$, $e_{u,\ v} = 1$; otherwise, $e_{u,\ v} = 0$.

3. *D2D Sharing Link*: On the basis of the physical connection $m_{u,v}^t$ and social relationship $e_{u,\ v}$, whether a V2V sharing link can be established between VUE u and VUE v at time slot t is judged by $a_{u,v}^t = m_{u,v}^t e_{u,v}$. In particular, $a_{u,v}^t = 1$ holds only if $m_{u,v}^t = 1$ and $e_{u,v} = 1$, i.e., the V2V sharing link can be successfully established only if the corresponding physical distance between the involved pair of VUEs is within the V2V communication range and if the two VUEs have a close social relationship. We define $\mathbf{A} = \left(a_{u,v}^t\right)_{U \times U}$ to denote the V2V sharing matrix at time slot t.

4.1.3 Content Transmission Latency Analysis

As shown in Fig. 4.4, there are two ways for VUEs to fetch their desired contents in this network. The corresponding latency of each way to fetch contents at time slot t is shown as follows:

Case 1: If the content f requested by VUE u is cached in the hired IV i, then a V2V link can be successfully established between VUE u and IV i, and VUE u can fetch the desired content f from IV i. In this case, the corresponding latency to fetch the content f can be expressed as $a_{i,u} x_{i,f} \tau_{i,u}^{f,t}$. Additionally, if VUE u is one of the hired IVs and the desired content f is cached in the caching list, the corresponding latency to fetch the content f is zero.

Case 2: If the content request cannot be satisfied by the hired IV i, VUE u sends its request to the RSU for the desired content f. In this case, the latency of VUE u fetching the desired content f can be expressed as $\left(1 - a_{i,u}^t x_{i,f}^t\right) \tau_{s,u}^{f,t}$.

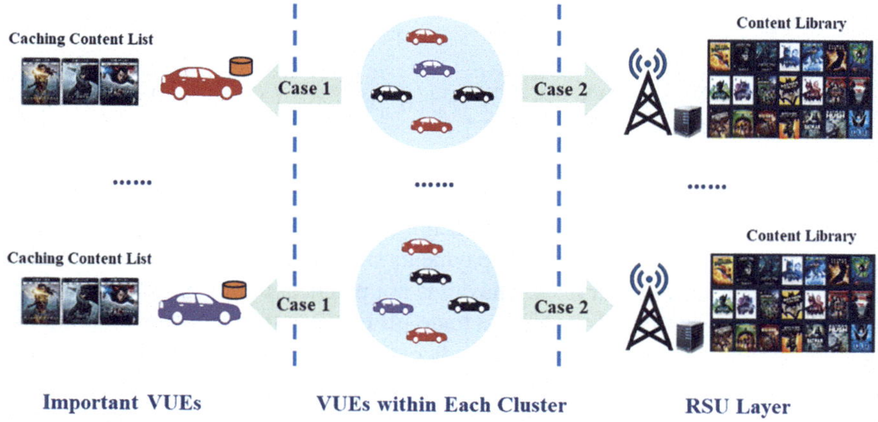

Fig. 4.4 Illustration of the content request path through different caching nodes

As a result, the total latency for all VUEs in the networks to fetch their desired contents at time slot t can be calculated as follows:

$$\tau_c = \sum_{i=1}^{I} \sum_{u=1}^{|C_i|} \sum_{f=1}^{F} r_u^{f,t} \left(a_{i,u}^t x_{i,f}^t \tau_{i,u}^{f,t} + \left(1 - a_{i,u}^t x_{i,f}^t\right) \tau_{s,u}^{f,t} \right), \tag{4.7}$$

where $| \, C_i \, |$ denotes the number of VUEs served by IV i at time slot t and where $r_u^{f,t}$ denotes the number of contents f requested by VUE u at time slot t.

4.1.4 Problem Formulation

From the operator's point of view, every penny it pays to hire IVs must bring the greatest benefit to the system. In this work, this benefit comes from the content transmission latency being reduced by hiring IVs, which is denoted by τ_b. According to the definition, τ_b can be expressed as:

$$\tau_b = \tau_{nc} - \tau_c, \tag{4.8}$$

where τ_{nc} represents the total latency of all VUEs to fetch their desired contents when no VUEs are hired by the operators in the network. This can be written as:

$$\tau_{nc} = \sum_{u=1}^{U} \sum_{f=1}^{F} r_u^{f,t} \tau_{s,u}^{f,t}. \tag{4.9}$$

Then, we can calculate τ_b as follows:

$$\tau_b = \sum_{u=1}^{U}\sum_{f=1}^{F} r_u^{f,t}\tau_{s,u}^{f,t} - \sum_{i=1}^{I}\sum_{u=1}^{|C_i|}\sum_{f=1}^{F} r_u^{f,t}\left(a_{i,u}^{t}x_{i,f}^{t}\tau_{i,u}^{f,t} + \left(1 - a_{i,u}^{t}x_{i,f}^{t}\right)\tau_{s,u}^{f,t}\right). \quad (4.10)$$

Formula (4.10) shows that the operators can reduce the content transmission latency by hiring more IVs to provide caching services. However, caching contents consumes the storage of IVs. Higher payments should be paid for IVs to incentivize them to provide caching services. Thus, it is not practical to hire all VUEs as IVs to provide caching services. To achieve a trade-off between the payments of hiring IVs and the additional benefit τ_b, the system benefit per unit cost G is defined as a new utility function when there are I IUs hired. It can be expressed as:

$$G = \frac{\tau_b}{\sum_{u=1}^{U}\sum_{f=1}^{F} r_u^{f,t}} \times \frac{1}{\sum_{i=1}^{I} qS_i + q_{\text{const}}}, \quad (4.11)$$

where q represents the price per storage unit that the operators need to pay IVs, S_i is the total caching capacity of IV i, and q_{const} represents the constant payment for hiring IVs, such as electric charges.

With the objective of maximizing the system benefit per unit cost, the optimization problem is formulated as:

$$\begin{aligned}
\mathcal{P}_1 : \quad & \max_{\mathcal{I},\mathbf{X}} \{G\}, \\
s.t. \quad C_1 : \quad & 0 \le I \le U,\ \mathcal{I} \subseteq \mathcal{U}, \\
C_2 : \quad & \sum_{f=1}^{F} x_{i,f} l_f \le S_i, \forall i \in \mathcal{I}, \\
C_3 : \quad & x_{i,f} \in \{0,1\}, \forall i \in \mathcal{I}, \forall f \in \mathcal{F}.
\end{aligned} \quad (4.12)$$

where C_1 constrains the number of selected IVs, C_2 denotes the maximum caching capacity of the selected IU i, and C_3 guarantees that $x_{i,f}$ is a binary variable.

The problem $\mathcal{P}_1$ is a mixed-integer linear programming problem, which has been proven to be NP-hard. In particular, if the numbers of VUEs and contents are large in the system, the dimension of the solution space will be very high, and the corresponding problems are difficult to solve. To address this challenge, problem $\mathcal{P}_1$ is decomposed into two stages, i.e., caching node selection and caching content placement. By solving these two subproblems, near-optimal solutions to the optimized problem $\mathcal{P}_1$ can be obtained. In addition, considering that traditional optimization approaches normally have extremely high computational complexity, which limits their potential for application in real systems, intelligent learning-based methods are adopted to solve these subproblems. In Sect. 4.2, we propose a community detection-based IV selection method to select the optimal VUE for caching services. To place the popular content in these selected VUEs, an AWFL-based content popularity prediction framework is proposed in Sect. 4.3.

4.2 Community Detection-Based IV Selection Method

In this section, a community detection-based method is proposed to solve the IV selection issue. First, the number of IVs is determined according to the distribution of the node degree in the social network. Next, by jointly considering the user's mobility and social properties, a V2V sharing graph-based community detection algorithm is innovatively proposed to group all VUEs into different communities. Finally, betweenness centrality is adopted to characterize the importance of the VUE for IV selection in each community.

4.2.1 Determining the Number of IVs

As described in Sect. 4.1.2, a V2V communication link is successfully established only if the involved pair of VUEs has a direct social relationship. Therefore, to guarantee that the selected IVs can successfully establish V2V communication links with at least ψ VUEs, the number of neighbors of each IV in the social network should be larger than ψ. Note that throughout the paper, one VUE is called a neighbor of the other VUE if it has social relationships with it. In a social network, the distribution of the node degree, i.e., the number of neighboring nodes, decays according to the power law distribution given by:

$$p(\psi) = c \times \psi^{-\varphi}, \tag{4.13}$$

where $\sum_{\psi=0} c \times \psi^{-\varphi} = 1$ and where $p(\psi)$ is the probability that a randomly chosen node has ψ neighbors, where c is a social constant and where φ denotes the decay coefficient. Inspired by (4.13), we obtain the probability that a randomly chosen node has at least ψ neighbors in a social network with U VUEs as follows:

$$p(u \geq \psi) = \sum_{u=\psi}^{U-1} p(u). \tag{4.14}$$

Then, the number of VUEs that have at least ψ neighbors in the social network can be calculated by:

$$I_{\psi} = \left\lfloor U \times \sum_{u=\psi}^{U-1} p(u) \right\rfloor, \tag{4.15}$$

where $\lfloor \cdot \rfloor$ is the floor function, which retrieves the largest integer that is equal to or smaller than x. As a result, if the selected IVs can successfully establish V2V communication links with at least ψ VUEs, the number of IUs should be less than I_ψ. Since I_ψ is a smaller integer value, the optimal number of IVs I can be easily found via an iterative method.

4.2.2 V2V Sharing Graph-Based Community Detection Algorithm

Once the number of IVs I is determined, all UEs are grouped into I disjoint communities. Within one community, one target VUE is selected as the IV to receive the contents from the RSU first and then to deliver them to the other target VUEs in the same community via V2V links. To guarantee that the selected IVs can successfully establish V2V links with the other VUEs within their community, a V2V sharing graph-based community detection algorithm is proposed.

The whole process of the proposed V2V sharing graph-based community detection algorithm is shown in Fig. 4.5. First, the V2V sharing graph $\mathcal{G}_A = \{\mathcal{U}, \mathcal{E}_A\}$ is constructed from the V2V sharing matrix $\mathbf{A}$, where $\mathcal{U}$ is the vertex set of VUEs and where $\mathcal{E}_A = \left\{a^t_{u,v}, \forall u, v \in \mathcal{U}\right\}$ is the edge set. As described in Sect. 4.1.2, if and only if VUE u and VUE v have a close physical distance and social relationship, $a^t_{u,v} = 1$; otherwise, $a^t_{u,v} = 0$. On the basis of the V2V sharing graph $\mathcal{G}_A$, the community detection algorithm is adopted to partition the graph $\mathcal{G}_A$ into a set of disjoint communities. The goodness of clustering achieved by community detection is widely measured by modularity Q. Given the graph $\mathcal{G}_A$ and its adjacency matrix $\mathbf{A}$, the modularity Q is given by:

$$Q = \sum_{c \in C} \left[\frac{A_{u,v}}{2m} - \left(\frac{a_{C_i}}{2m} \right)^2 \right], \tag{4.16}$$

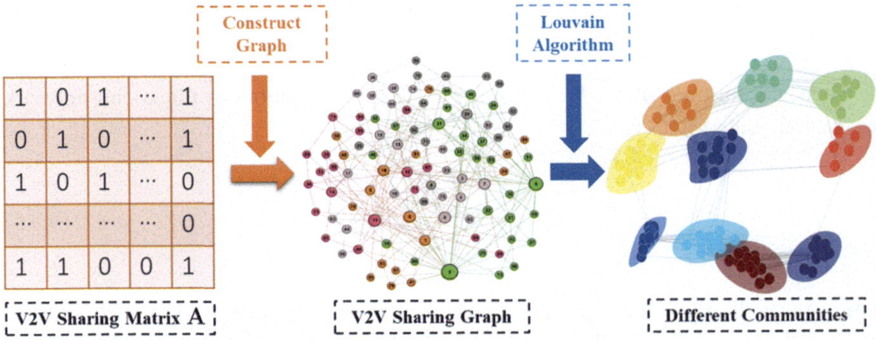

Fig. 4.5 Illustration of the V2V sharing graph-based community detection algorithm

where $m = \sum_{a(u,v) \in \mathcal{E}_A} a^t_{u,v}$ is the sum of all the edge weights, $A_{u,v} = \sum a^t_{u,v}, \forall u, v \in C_i$ denotes the sum of all edge weights between vertices within a particular community C_i, and a_{C_i} represents the sum of weights of all the edges incident upon each community C_i. However, modularity optimization is an NP-hard problem. Many methods use heuristics to find the best value of Q in NP-hard community detection problems. Among these methods, the Louvain algorithm outperforms other methods in terms of computational time, which is linear with respect to the number of edges in the graph, i.e., $O(|\mathcal{E}_{\mathcal{A}}|)$ [18].

The Louvain algorithm consists of multiple phases. Each phase runs for numerous iterations until reaching convergence. Initially, each vertex is assigned to a separate community. Within each iteration, all vertices are processed as follows: for a given vertex u, the gain in modularity ($\Delta Q_{u\to C_i}$) results in moving u to any of its neighboring communities, which can be calculated by

$$\begin{aligned}\Delta Q_{u\to C_i} &= \left[\frac{A_{u,v} + a_{u\to C_i}}{2m} - \left(\frac{a_{C_i} + a(u)}{2m}\right)^2\right] \\ &\quad - \left[\frac{A_{u,v}}{2m} - \left(\frac{a_{C_i}}{2m}\right)^2 - \left(\frac{a(u)}{2m}\right)^2\right], \\ &= \frac{a_{u\to C_i}}{2m} - \frac{a_{C_i} \times a(u)}{2m^2},\end{aligned} \tag{4.17}$$

where $a_{u\to C_i} = \sum_{v \in C_i} a_{u,v}$ is the sum of the weights of the edges from vertex u to all the vertices in community C_i and where $a(u)$ is the sum of the weights of all the edges incident to vertex u. If the maximum gain $\Delta Q_{u\to C_i}$ is positive, then u is moved to C_i from its current community. This continues until there is no positive modularity gain, u remains in that community, and the process ends. When a phase ends, the graph for the next phase is rebuilt by collapsing all vertices within a community into a single meta-vertex. The whole process is continued until all VUEs are grouped into communities $C = \{C_1, C_2, \ldots, C_i, \ldots, C_I\}$.

4.2.3 Selecting the IU for Each Community

After VUEs are grouped into I communities, a selection scheme is proposed to select the IV within each community. The betweenness centrality B_u and the available storage capacity S_u are defined to characterize the importance of the VUE u, which is denoted by V_u and is expressed as

$$V_u = \gamma \times B_u + (1 - \gamma) \times S_u, \tag{4.18}$$

where γ is a tunable parameter. The betweenness centrality B_u measures the social importance of user u. It can be calculated as

$$B_u = \sum_{j=1}^{U} \sum_{j<k} \frac{g_{j,k}(u)}{G_{j,k}}, \tag{4.19}$$

where $G_{j,\ k}$ is the number of shortest links between VUE j and VUE k and where $g_{j,\ k}(u)$ is the number of shortest links between VUE j and VUE k via VUE u. Then, the importance of different VUEs within each community can be obtained. Finally, the VUE with the top importance is chosen as the IV to cache files within each community. The entire process of the proposed IV selection algorithm is shown in Algorithm 4.1.

Algorithm 4.1: The Community Detection-Based IU Selection Algorithm

Input: D2D sharing matrix **A**, the minimum number of neighbors ψ, the social network $\mathcal{G}_S$, the caching capacity of each VUE S_u.

Determine the number of IVs I according to the distribution of the node degree;
Construct V2V sharing graph $\mathcal{G}_A = \{\mathcal{U}, \mathcal{E}_A\}$from the V2V sharing matrix **A**;
$\mathcal{C} \leftarrow$ Initialize each VUE in its own community;
While true **do**
  For all $u \in \mathcal{U}$**do**
    $\mathcal{N}(u) \leftarrow$ neighboring communities of u;
    Compute the gain in modularity $\Delta Q_{u \to c}$;
      Target community $\widehat{c} \leftarrow \underset{v \in \mathcal{N}(u)}{\text{argmax}} \Delta Q_{u \to c}$;
    If $\Delta Q_{u \to c} > 0$ **then**
      Move u to $\widehat{c}$ and update $\mathcal{C}$;
    Rebuild graph $\mathcal{G}_A = \{\mathcal{U}, \mathcal{E}_A\}$ by collapsing all vertices within a community into a single meta-vertex;
    Repeat until VUEs are grouped into communities $\mathcal{C} = \{\mathcal{C}_1, \mathcal{C}_2, \ldots \mathcal{C}_I\}$;
For each community $\mathcal{C}_i$ **do**
Calculate the importance of VUEs in $\mathcal{C}_i$ by $V_u = \gamma \times B_u + (1 - \gamma) \times S_u$;
$I_i \leftarrow$Sort the importance of VUEs and choose the top-1 as IV within each community;
Return the set of IVs $\mathcal{I} = \{I_1, I_2, \ldots, I_I\}$.
Output: The set of IVs $\mathcal{I}$.

4.3 Attention-Weighted Federated Learning-Based Content Popularity Prediction Method

4.3.1 *Attention-Weighted Federated Learning Architecture*

In view of user privacy concerns, FL is designed to collaboratively train a global ML model by using the local data at distributed VUEs. Compared with the centralized mode, it can significantly mitigate the privacy risk of vehicular users and largely reduce the communication costs. However, in complex and dynamic wireless environments, the heterogeneous activity level and computing capacity abilities of VUEs are challenging for a typical FL method. To address this challenge, an attention-weighted federated learning architecture is proposed, as shown in Fig. 4.6. Training a model in FL is performed through multiple communication rounds, and each communication round consists of the following steps.

1. *Local Training of UE Selection*: According to the selection ratio of clients, some VUEs are randomly selected to participate in FL training for the r th communication round. The set of the selected UEs is denoted as $\mathcal{K} = \{1, 2, \ldots, k, \ldots, K\} \subseteq \mathcal{U}$. The selection ratio of clients depends mainly on the channel condition since good channel conditions can guarantee that more VUEs upload the local trained model to the RSU server.
2. *Model Download*: Once the set of VUEs $\mathcal{K}$ is selected, the next step in the FL training process is to download the global model $\mathbf{w}_r$, $\mathbf{b}_r$, to the selected VUEs from the RSU server, as shown in Fig. 4.6, step (2). To reduce the consumption of wireless resources, the RSU can distribute the global model to these selected VUEs via broadcasting.
3. *Local Training*: After the selected VUEs have downloaded the global model $\mathbf{w}_r$, $\mathbf{b}_r$, the next step is to train the downloaded global model by utilizing the local data

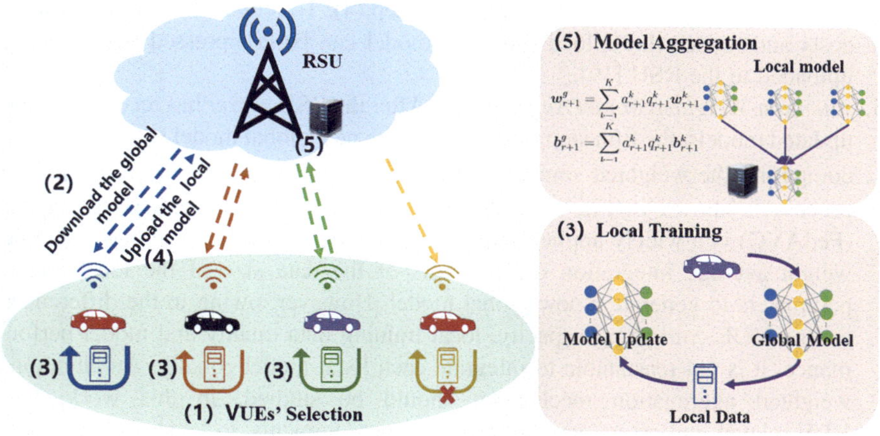

Fig. 4.6 Illustration of the attention-weighted federated learning architecture

at all the selected VUEs. The process is shown in Fig. 4.6, step (3). Let $\mathcal{H} = \{\mathcal{H}_1, \mathcal{H}_2, \mathcal{H}_3, \ldots, \mathcal{H}_k, \ldots, \mathcal{H}_K\}$ represent the local datasets stored in the selected VUEs. $\mathcal{H}_k$ represents the local dataset of VUE k with length n_k, where $n_k = \mid \mathcal{H}_k|$. The goal of local training at the selected VUE k is to minimize the loss between the prediction and the ground truth by utilizing local data $\mathcal{H}_k$. The loss function of local training at the selected VUE k can be defined as

$$\min_{\mathbf{w},\mathbf{b}} \quad F\left(\mathbf{w^k}, \mathbf{b^k}\right) = \frac{1}{n_k} \sum_{i=1}^{n_k} f(\mathbf{w}_r, \mathbf{b}_r; s_i), \tag{4.20}$$

$$f(\mathbf{w}_r, \mathbf{b}_r; s_i) = \left\| Y_i - \widehat{Y}_i \right\|, \tag{4.21}$$

where $\mathbf{w^k}$ and $\mathbf{b^k}$ are the weights and biases of the local trained model, respectively. s_i denotes one sample to train the model at the selected VUE k, where $s_i \in \mathcal{H}_k$. Y_i and $\widehat{Y}_i$ are the ground truth and prediction values, respectively. The Adam algorithm is used to update the model parameters, which can be obtained as follows:

$$\mathbf{w}_{r+1}^k \leftarrow \mathbf{w}_r - \eta \sum_{i \in b} \nabla f(\mathbf{w}_r; s_i), \tag{4.22}$$

$$\mathbf{b}_{r+1}^k \leftarrow \mathbf{b}_r - \eta \sum_{i \in b} \nabla f(\mathbf{b}_r; s_i), \tag{4.23}$$

where $\nabla f(\mathbf{w}_r; s_i)$ and $\nabla f(\mathbf{b}_r; s_i)$ are the gradients with respect to $\mathbf{w}$ and $\mathbf{b}$, respectively, which can be obtained via backpropagation, and where η is the learning rate.

4. *Upload the Updated Model*: Once the local training at the selected VUE k is completed, the updated model $\mathbf{w}_{r+1}^k$, $\mathbf{b}_{r+1}^k$, is sent back to the RSU server for model aggregation, as shown in Fig. 4.6, step (4). To reduce the communication costs and reduce the upload time, the model can be compressed before being uploaded to the RSU [19].
5. *Attention-Weighted Model Aggregation*: After the RSU server has received all the updated models, the fifth step is to generate the new global model $\mathbf{w}_{r+1}$, $\mathbf{b}_{r+1}$, by computing the weighted sum of all the received local models $\mathbf{w}_{r+1}^k$, $\mathbf{b}_{r+1}^k$. The process is shown in Fig. 4.6, step (5). In typical FL, Federated Averaging (FedAVG) is a widely applied model aggregation algorithm. It employs simple weight average integration on the basis of the data size of the local model parameters to generate a new global model. However, owing to the differences among VUE computing capacity, local training data quality and model performance, it is not reasonable to integrate each local model equally. An attention-weighted aggregation mechanism should be studied. In this work, two VUE-related indicators are adopted as measurements to evaluate the local model's contributions to the global model. The process of the attention-weighted model aggregation is shown in Fig. 4.7.

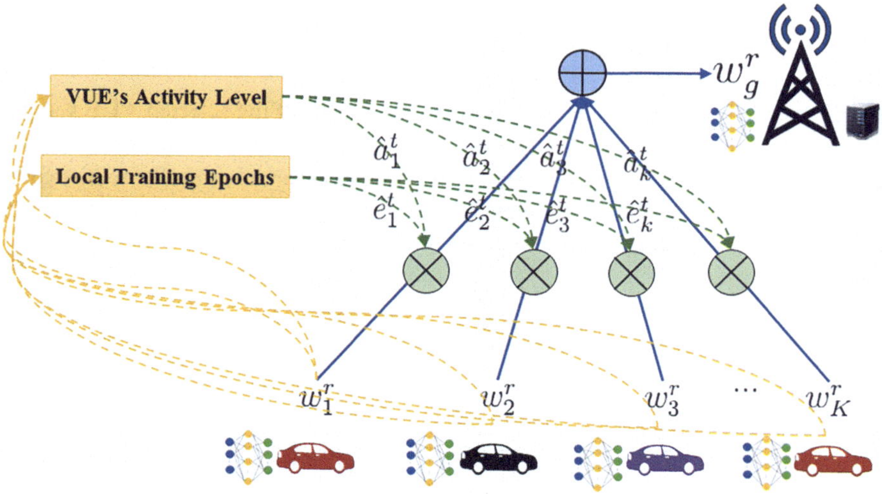

Fig. 4.7 Process of attention-weighted model aggregation

As shown in Fig. 4.7, the first indicator is the VUE activity level. In general, since young people use mobile phones more frequently, the activity level of teenagers is higher than that of middle-aged people. This implies that a VUE with high activity levels will have more local training data, which can train high-quality models. In other words, the contributions of VUEs to the global model differ greatly depending on activity level. Therefore, the historical numbers of requests within an observation window are utilized to estimate VUE activity levels in this work. The activity level of the VUE k can be expressed as

$$a_{r+1}^{k} = \frac{\left\| \mathbf{d}_{t-\rho:t}^{k} \right\|}{\sum_{k=1}^{K} \left\| \mathbf{d}_{t-\rho:t}^{k} \right\|}, k \in \{1, \ldots, K\}, \tag{4.24}$$

where $\left\| \mathbf{d}_{t-\rho:t}^{k} \right\|$ is the total number of requests of all contents uploaded by VUE k in the observation window $[t-\rho, t]$ and where ρ is the size of the observation window.

The second indicator is the number of local training epochs. The computing capacities of different VUEs are different, which results in varied times in computing the same training task. This also results in different numbers of local training epochs at different VUEs. As the number of training epochs is a key factor in determining model convergence, a larger weight should be assigned to local training models with a larger number of epochs. Since the model updates slow as the number of epochs increases, a log function is adopted to calculate the weights of the local training epochs. The weight of the selected VUE's k local training epochs can be calculated by

$$q_{r+1}^{k} = \frac{\log\left(\mathrm{e}_{\mathrm{r}+1}^{\mathrm{k}}\right)}{\log\left(\max\left\{\mathbf{e}_{\mathrm{r}+1}\right\}\right)}, \mathbf{e}_{r+1} = \left\{e_{r+1}^{1}, e_{r+1}^{2}, \ldots, e_{r+1}^{K}\right\}, \tag{4.25}$$

where e_{r+1}^{k} denotes the number of local training epochs of the selected VUE k at the $(r + 1)$ th communication round. Notably, the maximum values of the local training epochs $\max\{\mathbf{e}_{\mathrm{r}+1}\}$ are chosen on the basis of the experimental results in this work.

According to the attention-weighted aggregation mechanism, the RSU server aggregates the local trained models of all VUEs by

$$\mathbf{w}_{r+1}^{g} = \sum_{k=1}^{K} a_{r+1}^{k} q_{r+1}^{k} \mathbf{w}_{r+1}^{k}, \tag{4.26}$$

$$\mathbf{b}_{r+1}^{g} = \sum_{k=1}^{K} a_{r+1}^{k} q_{r+1}^{k} \mathbf{b}_{r+1}^{k}. \tag{4.27}$$

After one round of FL training, VUEs refresh their local model parameters to the latest downloaded parameters to prepare for the next round of FL training.

4.3.2 *BiLSTM-Based Content Popularity Prediction Model*

Owing to regular life patterns, most content request patterns are periodic in the time domain, such as 1 day or 1 week. In this work, BiLSTM (an extension of traditional LSTM) is adopted as the backbone for our deep learning (DL) model to predict the content popularity in each community. Compared with LSTM, BiLSTM can effectively increase the amount of available information to improve prediction accuracy. The framework of the proposed content popularity prediction model is shown in Fig. 4.8.

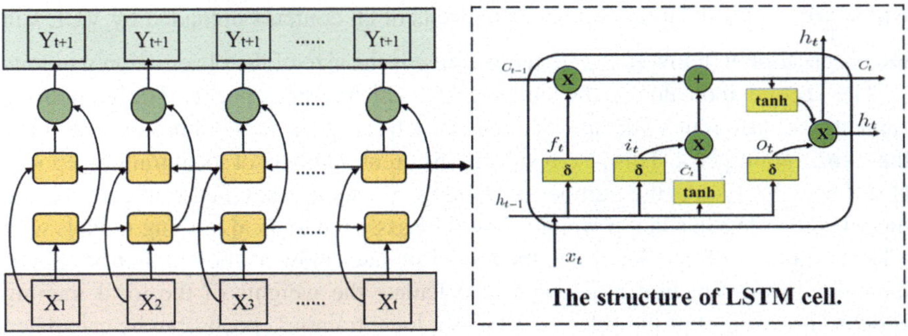

Fig. 4.8 Framework of the BiLSTM-based content popularity prediction model

In this model, the historical numbers of content requests in an observation window with a length of ρ time periods are utilized as inputs, which are denoted by $\mathbf{D} = [D_{t-\rho}, D_{t-\rho+1}, \ldots, D_t]$. After the historical data are input into the model, the memory of the historical information for long and short periods is controlled through three self-parameterized controlling gates, i.e., the input gate g_i^t, forget gate g_f^t and output gate g_o^t. The operations can be expressed as

$$\begin{aligned} g_i^t &= \delta(\mathbf{w}_i[h_{t-1}, D_t] + \mathbf{b}_i), \\ g_f^t &= \delta(\mathbf{w}_f[h_{t-1}, D_t] + \mathbf{b}_f), \\ g_o^t &= \delta(\mathbf{w}_o[h_{t-1}, D_t] + \mathbf{b}_o), \\ c_t &= g_f^t c_{t-1} + g_i^t \tanh(\mathbf{w}_c[h_{t-1}, D_t] + \mathbf{b}_c), \\ h_t &= g_o^t \tanh(\mathbf{c}_t), \end{aligned} \tag{4.28}$$

where $\mathbf{w}_i$, $\mathbf{b}_i$, $\mathbf{w}_f$, $\mathbf{b}_f$, $\mathbf{w}_o$, $\mathbf{b}_o$, $\mathbf{w}_c$ and $\mathbf{b}_c$ are the weights and biases of BiLSTM to be learned; δ denotes the activation function; and tanh represents the tanh activation function $\frac{e^x - e^{-x}}{e^x + e^{-x}}$. The fully connected network is then adopted to fuse the BiLSTM outputs, which can be expressed as $\mathbf{o} = h_1 \oplus h_2 \cdots \oplus h_t$. Finally, after an activation operation, the request numbers of the content f at the $(t + 1)$ th time slot can be obtained by

$$\widehat{Y}_{t+1} = \delta(\mathbf{o}). \tag{4.29}$$

4.3.3 *Proactive Edge Caching Placement Policy*

As described in Section II-B, in this work, we assume that the VUEs in the same community have similar user preferences. Since each IV is a VUE in the same community, the content popularity predicted by the IV can represent the content popularity of the corresponding community. Thus, after all the communication rounds are completed, each IV i downloads the aggregated global models $\mathbf{w}_{r+1}$ and $\mathbf{b}_{r+1}$ from the RSU server. Then, the request numbers of the content f at the $(t + 1)$ th time slot are predicted, which is denoted by $\widehat{Y}_{t+1}^{i,f}$. Next, the popularity of the content f at the $(t + 1)$ th time slot can be written as

$$p_{t+1}^{i,f} = \frac{\widehat{Y}_{t+1}^{i,f}}{\sum_{f=1}^{F} \widehat{Y}_{t+1}^{i,f}}. \tag{4.30}$$

According to the vector of the content popularity $\mathbf{p}_{t+1}^{i} = \left[p_{t+1}^{i,1}, p_{t+1}^{i,2}, \ldots, p_{t+1}^{i,f}, \ldots, p_{t+1}^{i,F}\right]$, the most popular S_i content during the $(t + 1)$ th time slot in community C_i can be obtained via the sorting algorithm. Then, the caching strategy $\mathbf{x}_i^{t+1} = \left[x_{i,1}^{t+1}, x_{i,2}^{t+1}, \ldots, x_{i,f}^{t+1}, \ldots, x_{i,F}^{t+1}\right]$ can be determined by setting the values corresponding to the most popular S_i contents to 1 and the others to 0. Finally, each IV downloads the most popular C_i contents from the RSU to update the caching list.

4.4 Simulations and Discussion

To evaluate the performance of the proposed CAFLPC strategy, extensive experiments are conducted on a real-world dataset. In this section, we first describe the dataset utilized. Then, we list the parameter settings in detail. We also introduce the baseline schemes and performance metrics for performance comparisons. To set the learning rate and maximum training epochs for these experiments, we analyze the effects of different hyperparameters. Finally, we show the performance of the proposed AWFL, AWFL_BiLSTM and CAFLPC strategies.

4.4.1 Dataset Description

The dataset utilized in this work consists of YouTube videos from Kaggle [20]. Every hour, real-time count observations (views, comments, likes, and dislikes) of 1600 videos were recorded in May 2018. The video list was retrieved on 7 May 2018, and the total number of time slots was 643 h.

In addition, in the real-world dataset, only the number of content requests is recorded on a per-video basis, and there is no user information. As we need the content request data of different VUEs to validate the proposed scheme, we map the request data in the dataset randomly to each of the VUEs to construct a dataset with user content request information. First, we adopt a random distribution to generate a value between 0 and 1 for each VUE, which represents the proportion of the content request number of each VUE to the total content request number in the dataset. Then, to ensure that the cumulative sum of the proportions of different VUEs is 1, we conduct a normalization process, i.e., the proportion of each user is divided by the cumulative sum of the proportions of all VUEs. Finally, according to the normalized proportion of each VUE, content request data in the dataset are mapped to different VUEs to obtain the content request data of different VUEs. The obtained content request data of each VUE can be expressed as

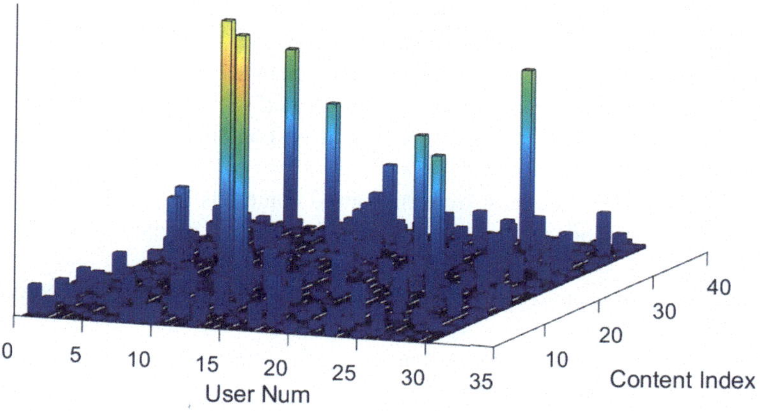

Fig. 4.9 Content preferences of different users

$$\mathbf{D}_u = \begin{bmatrix} r_u^{1,1} & \cdots & r_u^{1,t} & \cdots & r_u^{1,T} \\ \vdots & \ddots & \vdots & \ddots & \vdots \\ r_u^{f,1} & \cdots & r_u^{f,t} & \cdots & r_u^{f,T} \\ \vdots & \ddots & \vdots & \ddots & \vdots \\ r_u^{F,1} & \cdots & r_u^{F,t} & \cdots & r_u^{F,T} \end{bmatrix}, \tag{4.31}$$

where the obtained number of contents F is 100 and the total number of time slots T is 643. For each UE, the first 70% of the time slot data of each VUE are utilized as the training dataset, the next 15% of the time slot data are used as the validation dataset, and the remaining 15% of the time slot data are used as the testing dataset. To better show the obtained content request data of each UE, we plot the content requests of partial UEs. The results are shown in Fig. 4.9.

4.4.2 Baseline Schemes and Performance Metrics

To evaluate the performance of the proposed AWFL framework and CAFLPC strategy, eleven baseline schemes are adopted and compared:

CL_BiLSTM: A method in which the BiLSTM-based content popularity prediction model is trained via centralized learning.

C_BiLSTM: A BiLSTM-based content popularity prediction model that is trained locally in each community and has no global model aggregation.

FedAvg_DNN: A local method with which the DNN-based prediction model is trained in each community and aggregated with the FedAvg algorithm.

AWFL_DNN: A method different from AWFL_BiLSTM, in which a DNN is adopted to predict the content popularity in each community.

FedAvg_BiLSTM: A BiLSTM-based content popularity prediction model that is trained locally in each community and aggregated with the FedAvg algorithm.

LRU: A method in which the least recently used content is replaced by the new content.

LFU: A method where the least frequently used content is replaced first.

CFLPC: This method aggregates the locally trained prediction models with FedAvg.

CAFLPC_NC: A caching strategy that does not group VUEs via community detection.

CAFLPC_NI: A caching strategy that randomly selects IVs in each community without considering the importance of VUEs.

Optimal: A caching strategy designed on the basis of prior knowledge of future content requests.

To evaluate the accuracy of the content popularity prediction, the following three performance metrics are adopted: (1) the MAE, which measures the absolute errors between the content popularity prediction and the ground truth; (2) the RMSE, which measures the differences between the content popularity prediction and the ground truth; and (3) the R^2, which measures a statistical metric that represents the proportion of the variance in the dependent variable that is predictable from the independent variable. For the RMSE and MAE, a smaller value indicates better performance. In contrast, for R^2, a larger value implies a better fit to the data and thus better performance. In addition, to evaluate the performance of different caching strategies, the following metrics are also adopted: (1) the average downloading latency and (2) the system benefit per unit cost.

4.4.3 Convergence Performance of the Proposed AWFL

The convergence of the AWFL_BiLSTM method is investigated, and comparisons are made with FedAvg_BiLSTM and CL_BiLSTM for different selection ratios of clients. The results are shown in Fig. 4.10, from which it can be seen that the prediction losses of the three methods all gradually converge as the number of communication rounds increases. Moreover, the proposed AWFL_BiLSTM method outperforms FedAvg_BiLSTM in terms of convergence speed and prediction accuracy for all the simulated selection ratios of clients. Compared with FedAvg_BiLSTM, the gain brought by the proposed AWFL_BiLSTM is up to 0.68%, 1.19%, 1.97%, 1.35%, 1.16%, and 4.19% in terms of the MAE when the selection ratios of the clients are 10%, 30%, 50%, 70%, 90%, and 100%, respectively. This illustrates that the proposed AWFL framework has a faster convergence speed and higher prediction accuracy than the classic FedAvg algorithm does. Moreover, Fig. 4.10f shows that AWFL_BiLSTM can achieve similar prediction

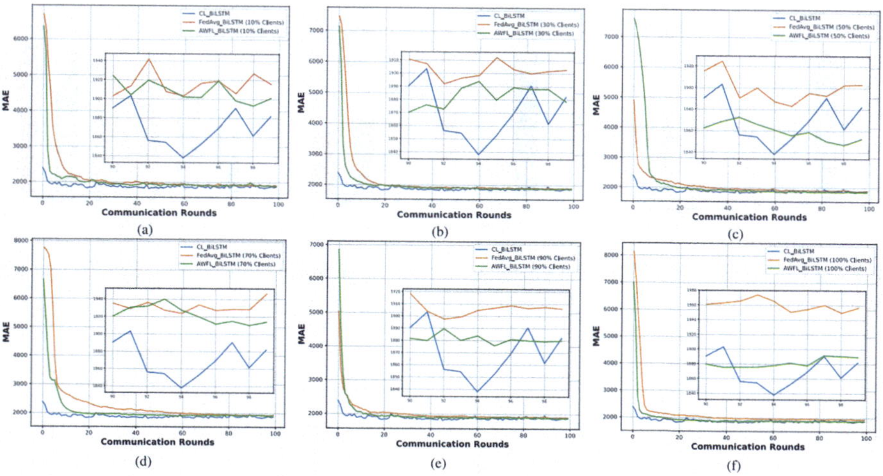

Fig. 4.10 Convergence of the proposed AWFL framework versus different selection ratios of clients

accuracy as CL_BiLSTM but is more stable. The reason behind this is that FL is a distributed framework where the data utilized to train the local model on each VUE are relatively smaller than those of CL and where the content preference of each VUE changes slowly over a relatively long period. This results in the parameters of the local model trained on each VUE in each communication round not being significantly updated. Therefore, compared with the global model trained by CL, the global model aggregated by FL is more stable.

4.4.4 Prediction Performance of the Proposed AWFL_BiLSTM

Experiments are conducted on a real-world dataset to verify the performance of AWFL_BiLSTM in terms of content popularity prediction. The obtained results are shown in Fig. 4.11. AWFL_BiLSTM achieves the best performance in terms of the MAE, RMSE and R^2 among all the conducted distributed learning methods. Compared with C_BiLSTM, FedAvg_BiLSTM and AWFL_BiLSTM perform better. This finding demonstrates that using the FL framework to aggregate the global model can improve the accuracy of the prediction. For the prediction methods based on the DNN, AWFL_DNN and FedAvg_DNN, the performance is not as good as that of AWFL_BiLSTM and FedAvg_BiLSTM. This is because BiLSTM networks have a better ability to capture the time dimension features of the content request patterns than DNN networks do. In addition, as shown in Fig. 4.11, the performance of the FedAvg_BiLSTM and FedAvg_DNN methods is inferior to that of AWFL_BiLSTM and AWFL_DNN. This finding validates our finding that the

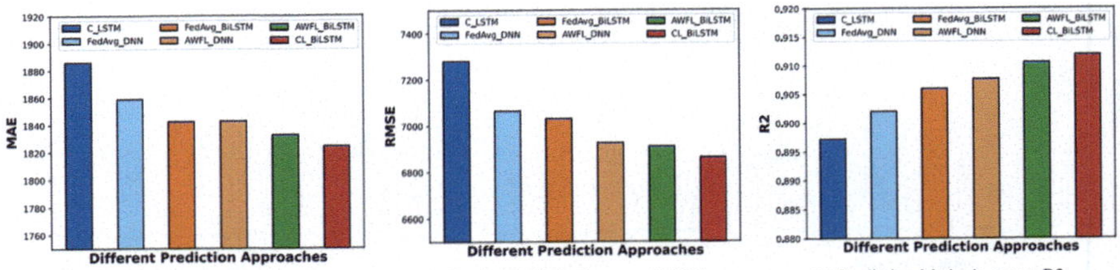

(a) Prediction Methods versus MAE (b) Prediction Methods versus RMSE (c) Prediction Methods versus R2

Fig. 4.11 Comparisons of different prediction algorithms in terms of the MAEs, RMSEs and R2 values

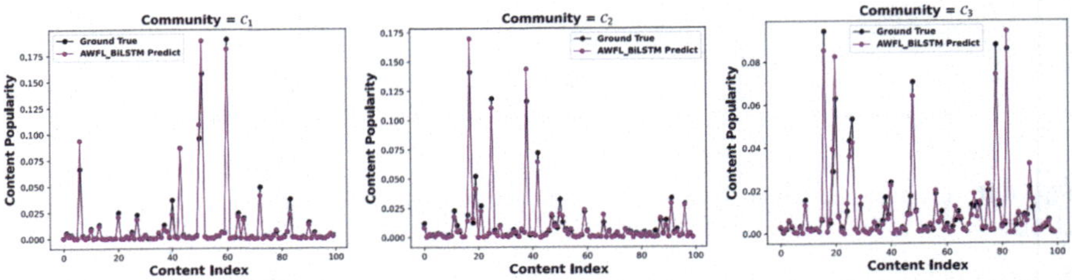

(a) Content Popularity Predicted in C_1 (b) Content Popularity Predicted in C_2 (c) Content Popularity Predicted in C_3

Fig. 4.12 The content popularity predicted by the proposed AWFL_BiLSTM

proposed AWFL framework can improve the prediction accuracy over that of the FedAvg algorithm. To further demonstrate the prediction performance of the proposed AWFL_BiLSTM method, the content popularity predicted by AWFL_BiLSTM is compared with the ground truth for different communities, and the results are shown in Fig. 4.12. The content popularity is highly dynamic in different communities, but the content popularity predicted by AWFL_BiLSTM fits the ground truth well. This demonstrates the effectiveness of AWFL_BiLSTM in content popularity prediction.

4.4.5 Caching Performance of the Proposed CAFLPC Strategy

To reveal the superiority of the proposed community detection-based IV selection method, the CAFLPC_NC and CAFLPC_NS strategies are simulated as baselines. LFU, LRU, and CFLPC are also adopted as baselines for evaluating the performance of the proposed CAFLPC strategy. The obtained results for different caching sizes, numbers of contents, and numbers of VUEs are shown in Fig. 4.13.

Figure 4.13a–d shows the performance comparisons of different caching strategies under different caching sizes. In these experiments, the maximum value of the VUE caching size ranges from 10 GB to 30 GB. Figure 4.13a, b shows that CAFLPC achieves the best performance in terms of the average downloading latency and

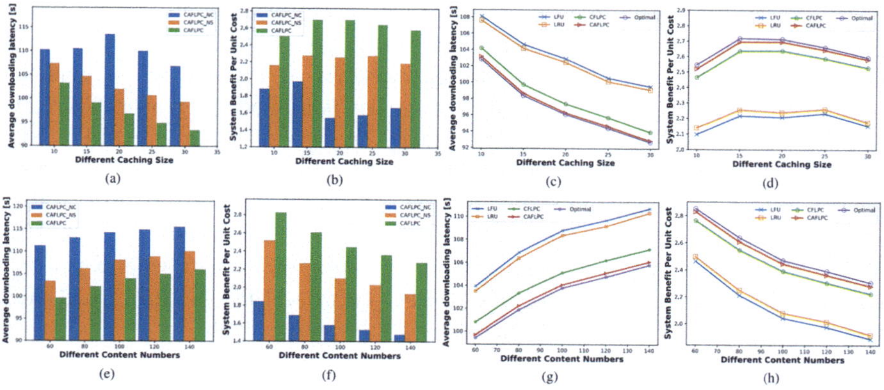

Fig. 4.13 Performance of different caching strategies versus different caching sizes and numbers of contents

system benefit per unit cost. Specifically, it outperforms CAFLPC_NC by up to 16.3% in average downloading latency and 42.9% in system benefit per unit cost when the caching size is 20 GB. The reason behind this is that the proposed community detection algorithm enables more UEs to obtain the desired contents from IUs via D2D links rather than from the BS and thus reduces the content downloading latency. Compared with those of CAFLPC_NS, the average downloading latency and system benefit per unit cost gain of CAFLPC reach 6.2% and 19.4%, respectively. The reason is that the IU selected by the CAFLPC can establish successful V2V communication links with multiple VUEs within a community. Figure 4.13c shows that the average downloading latency of all the strategies decreases as the caching size increases, since more content can be downloaded and stored. The average downloading latency gains brought by CAFLPC over LFU, LRU, and CFLPC are 6.9%, 6.4%, and 1.2%, respectively. As shown in Fig. 4.13d, the system benefit per unit cost decreases after the caching size reaches 15 GB, since more payments should be paid to the IU for the caching service as the caching size increases. The system benefits per unit cost gains of CAFLPC over LFU, LRU, and CFLPC are 25.9%, 24.7%, and 2.2%, respectively. In addition, the proposed CAFLPC always outperforms CFLPC in all the simulated performance metrics. This is because the quality of those local models is different, and aggregating them with the FedAvg algorithm cannot produce a well-performing global model. In addition, Fig. 4.13c, d shows that the content downloading latency decreases as the caching size increases, but the system benefit per unit cost increases and then decreases. This demonstrates that the average downloading latency can be reduced by increasing the caching size, but the system benefit per unit cost may not increase since the payments needed to be paid for providing caching services may increase. This is the reason we adopt maximizing the system benefit per unit cost as the goal of this work to achieve a trade-off between the average downloading latency and the payments.

Figure 4.13e–h shows the performance comparisons of different caching strategies for different numbers of contents. Figure 4.13e, g show that the average content downloading latency of all the strategies increases as the number of contents increases. This is because the caching size of IUs is limited. Moreover, CAFLPC achieves the lowest downloading latency compared with CAFLPC_NC, CAFLPC_NS, LFU, LRU, and CAFLPC. Specifically, the average downloading latency of CAFLPC outperforms that of CAFLPC_NC, CAFLPC_NS, LFU, LRU, and CAFLPC by 10.6%, 3.9%, 4.6%, 4.1%, and 1.1%, respectively, when the number of contents is 120. As shown in Fig. 4.13f, h, the system benefits per unit cost of CAFLPC outperform those of CAFLPC_NC, CAFLPC_NS, LFU, LRU and CFLPC at 35.1%, 15.1%, 21.5%, 20.0%, and 2.4%, respectively.

4.5 Summary

This work investigated the proactive edge caching strategy in V2V-assisted VSNs to achieve a trade-off between the cost for the system to pay for providing caching services and the content transmission latency reduced by hiring IVs. In doing so, an optimization problem is formulated, in which two challenges are solved: (1) which VUEs should be selected as caching nodes to provide caching services and (2) how to place the popular content on those selected VUEs without violating user privacy. To solve these problems, a CAFLPC strategy was proposed. Specifically, considering user mobility and social properties, a community detection-based IV selection method was proposed to select the optimal VUEs as IVs to provide caching services. To place the proper popular content in these selected IUs, an AWFL-based content popularity prediction framework was established. Most importantly, considering the imbalance of VUE activity levels and local computing capacities, an attention-weighted aggregation mechanism was proposed to improve the training efficiency and prediction accuracy. Intensive simulations were performed, and the results verified that the proposed CAFLPC strategy outperforms the existing caching strategies. Notably, on the basis of this work, a joint allocation of communication and caching resources can be performed to further reduce the content downloading latency.

References

1. Sun, R., Wang, Y., Lyu, L., Cheng, N., Zhang, S., Yang, T., & Shen, X. (2020). Delay-oriented caching strategies in D2D mobile networks. *IEEE Transactions on Vehicular Technology, 69*(8), 8529–8541.
2. Noor-A-Rahim, M., Liu, Z., Lee, H., Ali, G. G. M. N., Pesch, D., & Xiao, P. (2022). A survey on resource allocation in vehicular networks. *IEEE Transactions on Intelligent Transportation Systems, 23*(2), 701–721.

3. Jiang, Y., Ma, M., Bennis, M., Zheng, F.-C., & You, X. (2019). User preference learning-based edge caching for fog radio access network. *IEEE Transactions on Communications, 67*(2), 1268–1283.
4. Song, C., Xu, W., Wu, T., Yu, S., Zeng, P., & Zhang, N. (2021). QoE-driven edge caching in vehicle networks based on deep reinforcement learning. *IEEE Transactions on Vehicular Technology, 70*(6), 5286–5295.
5. Li, Z., Yang, C., Huang, X., Zeng, W., & Xie, S. (2023). CoOR: Collaborative task offloading and service caching replacement for vehicular edge computing networks. *IEEE Transactions on Vehicular Technology, 72*(7), 9676–9681.
6. Wu, H., Zhang, J., Cai, Z., Liu, F., Li, Y., & Liu, A. (2020). Toward energy-aware caching for intelligent connected vehicles. *IEEE Internet of Things Journal, 7*(9), 8157–8166.
7. Zhou, H., Jiang, K., He, S., Min, G., & Wu, J. (2023). Distributed deep multi-agent reinforcement learning for cooperative edge caching in internet-of-vehicles. *IEEE Transactions on Wireless Communications, 22*(12), 9595–9609.
8. Yao, L., Wang, Y., Wang, X., & Wu, G. (2021). Cooperative caching in vehicular content centric network based on social attributes and mobility. *IEEE Transactions on Mobile Computing, 20*(2), 391–402.
9. Aung, N., et al. (2023). VeSoNet: Traffic-aware content caching for vehicular social networks using deep reinforcement learning. *IEEE Transactions on Intelligent Transportation Systems, 24*(8), 8638–8649.
10. Wu, H., Fan, Y., Jin, J., Ma, H., & Xing, L. (2023). Social-aware decentralized cooperative caching for internet of vehicles. *IEEE Internet of Things Journal, 10*(16), 14834–14845.
11. Wang, R., Kan, Z., Cui, Y., Wu, D., & Zhen, Y. (2021). Cooperative caching strategy with content request prediction in internet of vehicles. *IEEE Internet of Things Journal, 8*(11), 8964–8975.
12. Yao, L., Xu, X., Deng, J., Wu, G., & Li, Z. (2022). A cooperative caching scheme for VCCN with mobility prediction and consistent hashing. *IEEE Transactions on Intelligent Transportation Systems, 23*(11), 20230–20242.
13. Qu, B., Li, D., & Li, X. (2021). Optimizing dynamic cache allocation in vehicular edge networks: A method combining multi-source data prediction and deep reinforcement learning. *IEEE Internet of Things Journal*. https://doi.org/10.1109/JIOT.2023.3324381
14. Jiang, W., Feng, G., Qin, S., Yum, T. S. P., & Cao, G. (2019). Multi-agent reinforcement learning for efficient content caching in Mobile D2D networks. *IEEE Transactions on Wireless Communications, 18*(3), 1610–1622.
15. Passarella, A., & Conti, M. (2013). Analysis of individual pair and aggregate Intercontact times in heterogeneous opportunistic networks. *IEEE Transactions on Mobile Computing, 12*(12), 2483–2495.
16. Zhang, W., Wu, D., Yang, W., & Cai, Y. (2019). Caching on the move: A user interest-driven caching strategy for D2D content sharing. *IEEE Transactions on Vehicular Technology, 68*(3), 2958–2971.
17. Wu, L., et al. (2017). Modeling the evolution of users' preferences and social links in social networking services. *IEEE Transactions on Knowledge and Data Engineering, 29*(6), 1240–1253.
18. Javed, M. A., Younis, M. S., Latif, S., et al. (2018). Community detection in networks: A multidisciplinary review. *Journal of Network and Computer Applications, 108*, 87–111.
19. Lv, S., Guo, S., & Zhang, H. (2021). Distribution-aware weight compression for federated averaging learning over wireless edge networks. In *2021 IEEE/CIC International Conference on Communications in China (ICCC), Xiamen, China* (pp. 1107–1112).
20. KMMD. (2018). *YouTube videos Viewcount every hour* [Online]. Available: https://www.kaggle.com/nnqkfdjq/statistics-observation-ofrandom-youtube-video

Chapter 5
Joint Communication and Computational Resource Management in VSNs

The existing computation and communication (2C) optimization schemes for cellular-based vehicle-to-everything (C-V2X) networks assume that the systems work in overlay mode [1, 2]. In this mode, the vehicle users (VUs) and cellular users (CUs) use different bandwidth resources, resulting in low spectrum utilization. To improve spectrum utilization and energy efficiency (EE), the underlying C-V2X network has been conceived as a promising solution [3, 4]. Previous works have focused on communication resource optimization. On the other hand, existing 2C optimization schemes for C-V2X networks have considered only the physical domain, and no social trust information has been considered. Owing to the open nature of vehicular networks, untrusted vehicles widely exist [5]. When tasks are offloaded to untrusted vehicles, the reliability of the computed results significantly decreases. This may lead to a re-offloading process and consume more power, thus decreasing EE.

To address these two problems, we study a social-mobility-aware underlying C-V2X framework. The main challenge for designing such a scheme is that the co-channel interference is quite complex, and the 2C resources are highly coupled. In addition, the underlying C-V2X framework includes three types of task offloading modes, i.e., local offloading, vehicle-to-infrastructure (V2I) offloading, and vehicle-to-vehicle (V2V) offloading. Integrating social and mobility attributes to select the optimal task offloading decision is another challenge. To solve this problem, we fully integrate both social and mobile characteristics and design an iterative edge server selection, task splitting, spectrum assignment, and power allocation algorithm. The main contributions of this study are summarized below.

- This chapter establishes an innovative social-mobility-aware underlying C-V2X framework where task vehicular users (T-VUs) are enabled to reuse the spectrum resource with CUs. To improve the resource utilization, we assume that the T-VU offloads tasks to a service vehicular user (S-VU) and roadside unit (RSU) via nonorthogonal multiple access (NOMA). To our knowledge, this work is the first to investigate secure task offloading by integrating social and mobility attributes

H. Zhang et al., *Intelligent Resources Management for Vehicular Social Networks*, Wireless Networks, https://doi.org/10.1007/978-3-031-80169-3_5

in the underlying C-V2X network. It not only prevents task offloading to untrusted vehicles but also improves resource utilization and system EE.

- On the basis of the proposed framework, we formulate an EE maximization problem to assign the edge server, computational task, spectrum, and power resources simultaneously through joint optimization. The formulated problem is multivariate and nonconvex. To solve this problem, we first convert the original fractional programming problem into a subtractive-form problem and then decompose it into separate computation and communication resource allocation subproblems.
- To solve the computational resource allocation subproblem, which involves intricate types of task offloading and remains nonconvex, we fully integrate the social and mobility characteristics and design a heuristic algorithm to achieve edge server selection and task splitting. To address the complex co-channel interference in the second subproblem, we propose a tightening lower bound method and a KM algorithm to obtain the power allocation and spectrum assignment solutions.
- We conduct intensive simulations to evaluate the system EE performance of the proposed algorithm. The simulation results show that the proposed algorithm achieves a higher EE than the baselines under different parameters do, which verifies the importance of social mobility awareness.

The rest of this chapter is organized as follows. In Sect. 5.1, the framework of 2C-enabled VSNs is described. The EE maximization problem for 2C is formulated in this section. Section 5.3 proposes a nonorthogonal multiple access–joint communication and computation resource allocation algorithm (NOMA-JCCRAA). The simulation results are presented in Sect. 5.4. Finally, Sect. 5.5 concludes the whole work.

5.1 Framework of Communication–Computation-Enabled VSNs

5.1.1 Social-Mobility-Aware Model

This work studies a social-mobility-aware underlying C-V2X network that uses NOMA to satisfy the differentiated quality of service (QoS) requirements for each CU and T-VU, as shown in Fig. 5.1. In the physical domain, a macro base station (MBS) is deployed to support high-rate data transmission of U CUs indexed by $u \in \mathcal{U} = \{1, 2, \ldots, U\}$, and S RSUs indexed by $s \in \mathcal{S} = \{1, 2, \ldots, S\}$ are deployed to support the computationally intensive services of M T-VUs indexed by $m \in \mathcal{M} = \{1, 2, \ldots, M\}$. Each RSU is equipped with a mobile edge computing (MEC) server. Given the limited computational capability of T-VUs, we allow T-VUs to offload computational tasks to nearby RSUs through V2I links and idle S-VUs through V2V links. We assume that there are N idle S-VUs indexed by

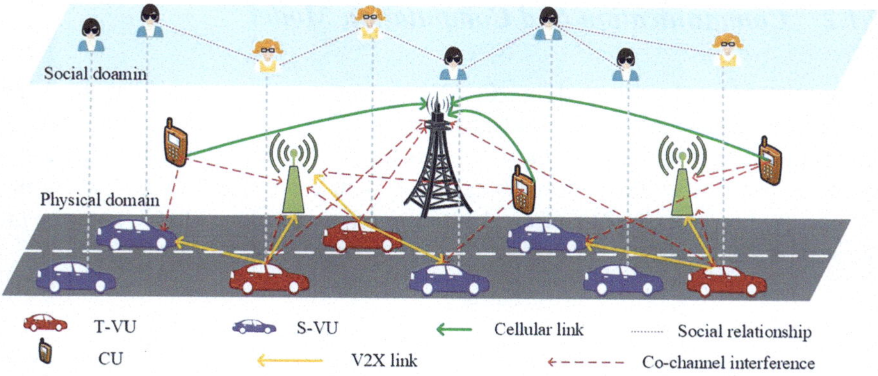

Fig. 5.1 System model for social-mobility-aware underlying C-V2X networks

$n \in \mathcal{N} = \{1, 2, \ldots, N\}$. On the basis of the characteristics of task offloading, this work allows the T-VU to offload tasks to the RSU server and the S-VU via NOMA.

In the social domain, leveraging social relationships can help build trustworthy V2V offloading links and improve the effective task offloading rate with limited power, i.e., EE. In this work, the social graph among VUs is denoted by $\mathcal{G} = (\mathbf{Z}, \boldsymbol{\delta})$, where $\mathbf{Z}$ denotes the set of all VUs with $\mathbf{Z} = \mathcal{M} \cup \mathcal{N}$ and where $\delta_{m,\,n} \in \boldsymbol{\delta} = \{\delta_{1,\,1}, \delta_{1,\,2}, \ldots, \delta_{M,\,N}\}$ represents the social trust relationship between the mth T-VU and the nth S-VU. According to [6], $\delta_{m,\,n}$ is defined as

$$\Psi_{m,v}(t) = k_{m,v}(t) \cdot \delta_{m,v}(t). \tag{5.1}$$

where $\zeta \in [0, 1]$ is the weight value of the direct trust and where $T_{di}^{m,n}$ and T_{in}^{n} denote the direct trust and indirect trust to the nth S-VU, respectively, which are defined as follows:

$$T_{di}^{m,n} = \frac{\alpha_1^{m,n}}{\left(\alpha_1^{m,n} + \alpha_2^{m,n}\right)}, \tag{5.2}$$

$$T_{in}^{n} = \sum_{n' \neq m} \theta_{n'} \frac{\alpha_1^{n,n'}}{\left(\alpha_1^{n,n'} + \alpha_2^{n,n'}\right)}, \tag{5.3}$$

in which α_1 and α_2 are the number of successful and unsuccessful offloaded tasks between vehicles in the historical statistics, respectively, and $\theta_{n'}$ is the credibility weight value of vehicle n' acting as an S-VU.

Since the primary innovation of this work is not to design a trustworthiness evaluation mechanism, we adopt a simple trustworthiness evaluation scheme. That is, if $\delta_{m,\,n} > \delta_{th}$, then vehicle n is trustworthy; otherwise, vehicle n is untrustworthy.

5.1.2 Communication and Computation Model

In the NOMA-enabled underlying C-V2X network, it is assumed that there are F available subchannels (SCs) indexed by $f \in \mathcal{F} = \{1, 2, \ldots, F\}$. To improve spectrum resource utilization, the CUs and the T-VUs are allowed to share the spectrum band. It is assumed that only one V2I link and one V2V link share the SC occupied by one CU. Therefore, the transmission rate of the uth CU over the fth SC at time slot t, $t \in \mathcal{T} = \{1, 2, \ldots, T\}$, can be expressed as

$$R_{u,f}^{[t]} = B \log_2 \left(1 + \frac{P_u^{op} X_{u,f}^{[t]} H_u^{[t]}}{\sum\limits_{m \in \mathcal{M}} \left(\epsilon_{m,1}^{[t]} + \epsilon_{m,2}^{[t]} \right) P_m^{th} X_{m,f}^{[t]} H_{m,u}^{[t]} + \sigma^2} \right), \tag{5.4}$$

where B represents the bandwidth of each SC; $\epsilon_{m,1}^{[t]}$ and $\epsilon_{m,2}^{[t]}$ are the power allocation coefficients from the mth T-VU to the RSU and to the S-VU at the tth time slot, respectively; P_m^{th} is the maximum transmit power of the mth T-VU; P_u^{op} denotes the optimal transmit power of the uth CU; σ^2 is the noise power; and the binary variable $X_{m,f}^{[t]} \in \{0, 1\}$ is defined as the spectrum assignment factor. If the mth T-VU occupies the fth SC at the tth time slot, $X_{m,f}^{[t]} = 1$; otherwise, $X_{m,f}^{[t]} = 0$. Similarly, $X_{u,f}^{[t]}$ is also a spectrum assignment indicator of the uth CU at the tth time slot. $H_u^{[t]}$ and $H_{m,u}^{[t]}$ are the channel and interference channel power gains of the uth CU at the tth time slot.

The signal-to-interference-plus-noise ratio (SINR) of the mth V2I link's receiver (i.e., the sth RSU) and the mth V2V link's receiver (i.e., the nth S-VU) can be separately expressed as

$$\gamma_{m,f}^{[t]} = \frac{\epsilon_{m,1}^{[t]} P_m^{th} X_{m,f}(t) H_{m,s}^{[t]}}{\sum\limits_{u \in \mathcal{U}} P_u^{op} X_{u,f}^{[t]} H_{u,s}^{[t]} + \epsilon_{m,2}^{[t]} P_m^{th} X_{m,f}^{[t]} H_{m,s}^{[t]} + \sigma^2}, \tag{5.5}$$

$$\gamma_{m,n,f}^{[t]} = \frac{X_{m,f}^{[t]} \Psi_{m,n}^{[t]} \epsilon_{m,2}^{[t]} P_m^{th} H_{m,n}^{[t]}}{\sum\limits_{u \in \mathcal{U}} P_u^{op} X_{u,f}^{[t]} \Psi_{m,n}^{[t]} H_{u,n}^{[t]} + \sigma^2}, \tag{5.6}$$

where $H_{m,s}^{[t]}$ and $H_{m,n}^{[t]}$ are the channel power gains from the mth T-VU to the sth RSU server and to the nth S-VU at the tth time slot, respectively; $H_{u,s}^{[t]}$ denotes the interference channel power gain from the uth CU to the sth RSU at the tth time slot; and $H_{u,s}^{[t]}$ is the interference channel power gain from the uth CU to the nth S-VU at the tth time slot. The binary variable $\Psi_{m,n}^{[t]}$, composed of both mobility and social relationships, is denoted as

$$\Psi_{m,n}^{[t]} = \begin{cases} 1, & \text{if } \rho_{m,n}^{[t]} < \rho_{th} \text{ and } \delta_{m,n}^{[t]} > \delta_{th}, \\ 0, & \text{otherwise,} \end{cases} \tag{5.7}$$

where ρ_{th} and δ_{th} are the thresholds of the physical and social domains, respectively, and $\rho_{m,n}^{[t]}$ is written as

$$\rho_{m,n}^{[t]} = \psi \cdot f\left(\Delta d_{m,n}^{[t]}\right) + (1 - \psi) \cdot f\left(\Delta v_{m,n}^{[t]}\right), \tag{5.8}$$

where $\Delta d_{m,n}^{[t]}$ is the distance between the T-VU and S-VU, $\Delta v_{m,n}^{[t]}$ represents the difference in velocity, $\psi \in [0, 1]$ is the weight of the distance, and $f(\cdot)$ is the normalized function.

We define a tuple $\left(D_m^{[t]}, C_m, \boldsymbol{\beta_m}^{[t]}\right)$ to characterize the task of the mth T-VU at the tth time slot, where $D_m^{[t]}$ is the size of the computation task and where C_m is the number of CPU cycles required for computing 1-bit data. We assume that each task is arbitrarily separable [7], and $\boldsymbol{\beta_m}^{[t]} = \left\{\beta_{m,\ 1}^{[t]}, \beta_{m,\ 2}^{[t]}\right\} \in [0,\ 1]$, where $\beta_{m,\ 1}^{[t]}$ represents the computing task splitting factor from the mth T-VU to the RSU server. $\beta_{m,\ 2}^{[t]}$ is the ratio computed by the S-VU. Thus, $\left(1 - \beta_{m,\ 1}^{[t]} - \beta_{m,\ 2}^{[t]}\right)$ denotes the portion of the computing task left for local execution (i.e., the mth T-VU). Therefore, the task execution delay at the mth T-VU is

$$\frac{D_m^{[t]}\left(1 - \beta_{m,1}^{[t]} - \beta_{m,2}^{[t]}\right) C_m}{y_m} < T_{\text{tol}}, \tag{5.9}$$

where y_m (in CPU cycle/s) is the assigned computing resource for executing local tasks and where T_{tol} denotes the maximum tolerance delay of each T-VU. The task offloading and executing delay from the mth T-VU to the RSU server and to the nth S-VU can be expressed as

$$\frac{D_m^{[t]} \beta_{m,1}^{[t]}}{R_{m,s,f}^{[t]}} + \frac{D_m^{[t]} \beta_{m,1}^{[t]} C_m}{y_{m,s}} < T_{\text{tol}}, \tag{5.10}$$

$$\frac{D_m^{[t]} \beta_{m,2}^{[t]}}{R_{m,n,f}^{[t]}} + \frac{D_m^{[t]} \beta_{m,2}^{[t]} C_m}{y_{m,n}} < T_{\text{tol}}, \tag{5.11}$$

where $R_{m,s,f}^{(t)} = B \log_2\left(1 + \gamma_{m,f}^{(t)}\right)$ denotes the transmission rate of the V2I link's receiver and where $R_{m,n,f}^{[t]} = B \log_2\left(1 + \gamma_{m,n,f}^{[t]}\right)$ is the transmission rate of the V2V link's receiver. $y_{m,\ s}$ (in CPU cycle/s) and $y_{m,\ n}$ (in CPU cycle/s) are the computing resources allocated to the mth T-VU served by the RSU server and the nth S-VU, respectively. The EE of the NOMA-enabled underlying C-V2X networks is expressed as

$$\xi = \sum_{t\in\mathcal{T}} \frac{R_{\text{total}}^{[t]}}{P_{\text{total}}^{[t]}} = \sum_{t\in\mathcal{T}} \sum_{n\in\mathcal{N}} \sum_{m\in\mathcal{M}} \sum_{s\in\mathcal{S}} \sum_{f\in\mathcal{F}} \frac{R_{m,s,f}^{[t]} + R_{m,n,f}^{[t]}}{\widetilde{P}_{m,n,s}^{[t]}}, \tag{5.12}$$

where $\widetilde{P}_{m,n,s}^{[t]} = \kappa y_m^3 + \left(\epsilon_{m,1}^{[t]} + \epsilon_{m,2}^{[t]}\right) P_m^{th} + \kappa y_{m,s}^3 + \kappa y_{m,n}^3$ and where κ is the effective switched capacitance depending on the CPU architecture.

5.2 Energy Efficiency Maximization Problem Formulation

In this work, our objective is to maximize the EE for the NOMA-enabled C-V2X network by optimizing edge server selection, task splitting, spectrum assignment and power allocation variables. Mathematically, the problem is formulated as

$$\mathcal{P}1: \max_{\Psi_{m,n}^{[t]}, \beta_{m,1}^{[t]}, \beta_{m,2}^{[t]}, X_{n,f}^{[t]}, X_{m,f}^{[t]}, \epsilon_{m,1}^{[t]}, \epsilon_{m,2}^{[t]}} \xi \tag{5.13a}$$

$$\text{s.t.}\, \Psi_{m,n}^{[t]}, X_{m,f}^{[t]}, X_{u,f}^{[t]} \in \{0,1\}, \forall m, s, n, u, f, t, \tag{5.13b}$$

$$0 \le \beta_m^{[t]} \le 1, \forall m, t, \tag{5.13c}$$

$$\epsilon_{m,1}^{[t]} \ge 0, \epsilon_{m,2}^{[t]} \ge 0, \epsilon_{m,1}^{[t]} + \epsilon_{m,2}^{[t]} \le 1, \forall m, t, \tag{5.13d}$$

$$R_{u,f}^{[t]} \ge R_{th,u}, \forall f, u, t, \tag{5.13e}$$

$$\sum_{f\in\mathcal{F}} X_{u,f}^{[t]} = \sum_{f\in\mathcal{F}} X_{m,f}^{[t]} = 1, \forall u, m, t, \tag{5.13f}$$

$$\sum_{u\in\mathcal{U}} X_{u,f}^{[t]} = 1, \quad \sum_{m\in\mathcal{M}} X_{m,f}^{[t]} \le 1, \quad \forall f, t, \tag{5.13g}$$

$$(5.9), (5.10), (5.11), \tag{5.13h}$$

where $R_{th,\,u}$ is the minimum data rate threshold for the uth CU, constraints (5.13c)-(5.13d) list the feasible task splitting and power allocation of the T-VUs, respectively, constraint (5.13e) represents the QoS requirements of CUs, and constraint (5.13f) restricts each user (T-VU and CU) to access only one SC; each SC can be shared by one CU and at most one T-VU according to constraint (5.13g).

It is obvious that (5.13a) represents fractional programming, which can be converted into a subtractive form. Therefore, (5.13a) is reformulated as

$$\max_{\Psi^{[t]}_{m,n},\beta^{[t]}_{m,1},\beta^{[t]}_{m,2},X^{[t]}_{m,f},X^{[t]}_{m,f},\epsilon^{[t]}_{m,1},\epsilon^{[t]}_{m,2}} \sum_{t\in\mathcal{T}}\left(R^{[t]}_{\text{total}}-\xi P^{[t]}_{\text{total}}\right). \tag{5.14}$$

5.3 Convex Optimization Theory-Based Communication and Computational Resource Management Strategy

Since the communication and computation resource decision of $\mathcal{P}1$ is made in each time slot and there is no interdependence among time slots, we transform the optimization problem across all the time slots into one time slot optimization problem. However, the obtained one-time slot optimization problem is still nonconvex, and it is difficult to obtain the global optimal solution. As an alternative, we decompose it into (1) a computational resource optimization subproblem $\mathcal{P}2$ and (2) a communication resource optimization subproblem $\mathcal{P}3$. $\mathcal{P}2$ and $\mathcal{P}3$ can be given by

$$\begin{aligned}\mathcal{P}_2:\ &\max_{\Psi^{[t]}_{m,n},\beta^{[t]}_{m,1},\beta^{[t]}_{m,2}} \left(R^{[t]}_{\text{total}}-\xi P^{[t]}_{\text{total}}\right)\\ \text{s.t.}\ &\Psi^{[t]}_{m,n}\in\{0,\ 1\},\ \forall m,n,\\ &(5.13\text{c}),(5.13\text{h}),\end{aligned} \tag{5.15}$$

$$\begin{aligned}\mathcal{P}3:\ &\max_{X^{[t]}_{u,f},X^{[t]}_{m,f},\epsilon^{[t]}_{m,1},\epsilon^{[t]}_{m,2}} \left(R^{[t]}_{\text{total}}-\xi P^{[t]}_{\text{total}}\right)\\ \text{s.t.}\ &X^{[t]}_{m,f},X^{[t]}_{u,f}\in\{0,\ 1\},\ \forall m,u,f,\\ &(5.10),(5.11),(5.13\text{d})-(5.13\text{g}).\end{aligned} \tag{5.16}$$

$\mathcal{P}2$ is a mixed-integer nonlinear programming (MINLP) problem. To find a tractable solution, we design a heuristic Social-Mobility-Aware Edge Server Selection and Task Splitting Algorithm (SM-SSTSA), as shown in Algorithm 5.1. To solve the communication resource allocation subproblem, we decouple $\mathcal{P}3$ into a power allocation subproblem and a spectrum assignment subproblem, which can be solved iteratively.

Algorithm 5.1: Social-Mobility-Aware Edge Server Selection and Task Splitting Algorithm (SM-SSTSA)

Initialize $\rho^{[t]}_{m,n}$, $\delta^{[t]}_{m,n}$.

FOR $m=1:M$ **DO**

Calculate $\rho^{[t]}_{m,n}$ based on (5.8).

Find the S-VU $n', n'\in\mathcal{N}$ that makes $\Psi^{[t]}_{m,n}=1$, and place it into the alternative S-VUs set $\mathcal{A}^{[t]}_m$.

Sort n' in a decreasing order according to the result of $R_{\text{total}}^{[t]} - \xi P_{\text{total}}^{[t]}$, and update the order of n' in $\mathcal{A}_m^{[t]}$.

FOR $i \in \mathcal{A}_m^{[t]}$ **DO**

Substitute i into (5.13h).

IF it exits $\beta_{m,1}^{[t]*}, \beta_{m,2}^{[t]*}$ satisfying (5.13h)

Break.

ENDIF

ENDFOR

$n^* = i$, $\Psi_{m,n^*}^{[t]} = 1$.

Delete n^* from the set $\mathcal{N}$.

ENDFOR

As proven in [8], when the spectrum assignment variable is fixed, (5.16a) is written as

$$
\begin{aligned}
& R_{m,s,f}^{[t]} + R_{m,n,f}^{[t]} - \xi P_{\text{total}}^{[t]} \\
\geq\ & \frac{\widetilde{\gamma}_{m,f}^{[t]}}{1+\widetilde{\gamma}_{m,f}^{[t]}} \log_2 \gamma_{m,f}^{[t]} + \log_2\left(1+\widetilde{\gamma}_{m,f}^{[t]}\right) - \frac{\widetilde{\gamma}_{m,f}^{[t]}}{1+\widetilde{\gamma}_{m,f}^{[t]}} \log_2 \widetilde{\gamma}_{m,f}^{[t]} \\
& + \frac{\widetilde{\gamma}_{m,n,f}^{[t]}}{1+\widetilde{\gamma}_{m,n,f}^{[t]}} \log_2 \gamma_{m,n,f}^{[t]} + \log_2\left(1+\widetilde{\gamma}_{m,n,f}^{[t]}\right) - \frac{\widetilde{\gamma}_{m,n,f}^{[t]}}{1+\widetilde{\gamma}_{m,n,f}^{[t]}} \log_2 \widetilde{\gamma}_{m,n,f}^{[t]} - \xi P_{\text{total}}^{[t]},
\end{aligned}
\tag{5.17}
$$

where $\widetilde{\gamma}_{m,f}^{[t]}$ and $\widetilde{\gamma}_{m,n,f}^{[t]}$ are the values of $\widetilde{\gamma}_{m,f}^{[t]}$ and $\widetilde{\gamma}_{m,n,f}^{[t]}$ from the previous iteration, respectively. We define $\Phi^{[t]}$ to represent the right-hand side of Ineq. (5.13a). Therefore, the lower bound of the objective function in (5.16a) can be written as $\max_{\epsilon} \Phi^{[t]}$. By denoting $\epsilon_{m,1}^{[t]} = 2^{w_{m,1}^{[t]}}$ and $\epsilon_{m,2}^{[t]} = 2^{w_{m,2}^{[t]}}$, the power control subproblem can be rewritten as

$$
\begin{aligned}
\mathcal{P}4: \quad & \max_{w_{m,1}^{[t]},\, w_{m,2}^{[t]}} \ \Phi^{[t]} \\
\text{s.t.} \quad & 2^{w_{m,1}^{[t]}} \geq 0, 2^{w_{m,2}^{[t]}} \geq 0, \ \ \forall m, \\
& 2^{w_{m,1}^{[t]}} + 2^{w_{m,2}^{[t]}} \leq 1, \ \ \forall m, \\
& (5.10), (5.11), (5.13e).
\end{aligned}
\tag{5.18}
$$

Since (5.18) is a standard convex optimization problem, we adopt Lagrangian dual decomposition to solve it. This solving process is summarized in Algorithm 5.2.

Given the power allocation decision, the spectrum assignment subproblem involves complicated matching among CUs, T-Vus, and SCs, which is proven to be NP-hard. From (5.13f)-(5.13g), the relationship between the CU and SC is a one-to-one match. To facilitate the solution, the complex match among CUs, T-Vus, and SCs is transformed into a new match among CUs and T-VUs. The new spectrum assignment variable between the uth CU and the mth T-VU at the tth time slot is denoted as $X_{u,m}^{[t]}$. Therefore, the spectrum assignment subproblem can be rewritten as

$$\begin{aligned} \mathcal{P}5: \quad & \max_{X_{u,m}^{[t]}} \left(R_{\text{total}}^{[t]} - \xi P_{\text{total}}^{[t]} \right) \\ \text{s.t.} \quad & X_{u,m}^{[t]} \in \{0,\ 1\},\ \forall u, m, \\ & \sum_{m \in \mathcal{M}} X_{u,m}^{[t]} \leq 1,\ \forall u, \\ & \sum_{u \in \mathcal{U}} X_{u,m}^{[t]} = 1,\ \forall m, \end{aligned} \tag{5.19}$$

which can be solved via a KM algorithm.

Algorithm 5.2: Power Allocation Algorithm for Each T-VU

Initialize the index $i = 0$, the power splitting $w_{m,1}^{[t]}\{i\}$, $w_{m,2}^{[t]}\{i\}$, the maximum number of iterations I_{ter}, and the convergence threshold ε.

  Calculate $\gamma_{m,f}^{[t]}\{i\}$ and $\gamma_{m,n,f}^{[t]}\{i\}$ via $w_{m,1}^{[t]}\{i\}$ and $w_{m,2}^{[t]}\{i\}$.

  WHILE $\alpha\{i\} \geq \varepsilon$ and $i < I_{\text{ter}}$

    $i = i + 1$.

    $\widetilde{\gamma}_{m,f}^{(t)}\{i\} = \gamma_{m,f}^{[t]}\{i-1\}$, $\widetilde{\gamma}_{m,n,f}^{[t]}\{i\} = \gamma_{m,n,f}^{[t]}\{i-1\}$.

    Solve $\mathcal{P}4$ utilizing Lagrange dual decomposition, and obtain $w_{m,1}^{[t]}\{i\}$, $w_{m,2}^{[t]}\{i\}$.

    Calculate $\gamma_{m,f}^{[t]}\{i\}$ and $\gamma_{m,n,f}^{[t]}\{i\}$.

    $\alpha\{i\} = \max\left\{ |\gamma_{m,f}^{[t]}\{i\} - \gamma_{m,f}^{[t]}\{i-1\}|,\ \gamma_{m,n,f}^{[t]}\{i\} - \gamma_{m,n,f}^{[t]}\{i-1\}| \right\}$.

  END WHILE

To solve $\mathcal{P}1$, the JCCRAA is proposed, which solves the computational resource allocation subproblem and the communication resource allocation subproblem. In the JCCRAA, by solving Algorithm 5.1, $\Psi_{m,n}^{[t]}$ and $\left\{ \beta_{m,1}^{[t]}, \beta_{m,2}^{[t]} \right\}$ can be obtained. Then, the analytical expressions of $\left\{ X_{u,f}^{[t]}, X_{m,f}^{[t]} \right\}$ and $\left\{ \epsilon_{m,1}^{[t]}, \epsilon_{m,2}^{[t]} \right\}$ can be derived via Algorithm 5.2 and the KM algorithm. Next, by substituting the obtained assigned spectrum and allocated power into Algorithm 5.1, the S-VU selection and task-splitting strategies are updated. The process is repeated until convergence, and the original problem is solved.

5.4 Simulations and Discussion

Intensive simulations are performed to evaluate the performance of the proposed algorithm. All the users are assumed to be located within a target rectangular area of 1000×1000 m. The simulation parameters are set according to 3GPP TR 36.885 [9], where an MBS is located at the center of the area and numerous RSUs are located at the roadside in the area. The number of lanes is 6, and the width of each lane is 4 m. The average inter-VU distance driving in the same lane is $2.5v$ m, with v representing the moving speed of vehicles in meters per second. In addition, we set $P_u^{op} = 20$ dBm, $P_m^{th} = [15, \; 30]$ dBm, and $D_m = [10^4, \; 10^5]$ bits.

The impacts of the number of T-VUs, the size of the offloaded tasks, the number of SCs and the number of untrusted S-VUs on the system EE are simulated. The obtained results are shown in Figs. 5.2, 5.3, 5.4, and 5.5. To demonstrate the superiority of the proposed algorithm, four baselines are simulated and compared:

1. NOMA-SMRAA, which is composed of a social-mobility-aware edge server selection scheme and a communication resource allocation algorithm.
2. NOMA-MTCRAA, which is composed of mobility-aware server selection and the proposed task-splitting and communication resource allocation algorithm.
3. RSU-SRAA, which is composed of the RSU-based offloading strategy and the proposed spectrum resource allocation algorithm.
4. Overlay-JCCRAA, which is adopted by the proposed JCCRAA and is based on the overlay mode.

The system EE for different numbers of T-VUs is shown in Fig. 5.2, from which we can see that the EE decreases as the number of T-VUs increases for all the simulated algorithms. The reason is that as the number of T-VUs increases, the competition for limited communication resources intensifies, resulting in severe co-channel interference and a degradation in EE performance. In addition, we see

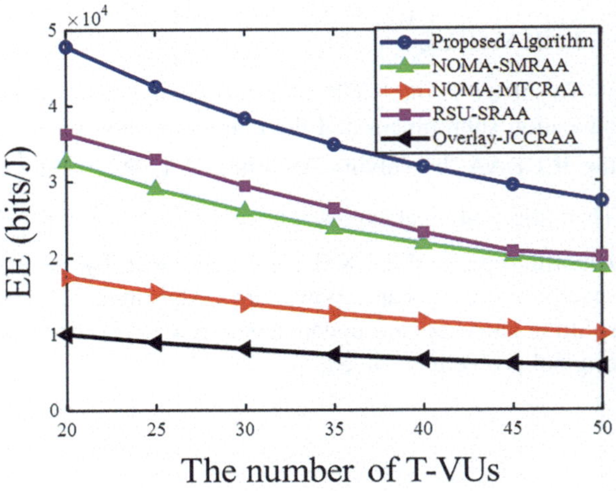

Fig. 5.2 System EE performance with respect to the number of T-VUs

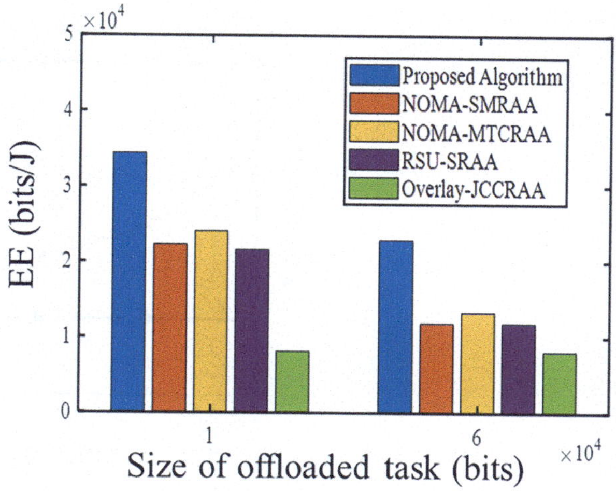

Fig. 5.3 System EE performance for different offloaded task sizes

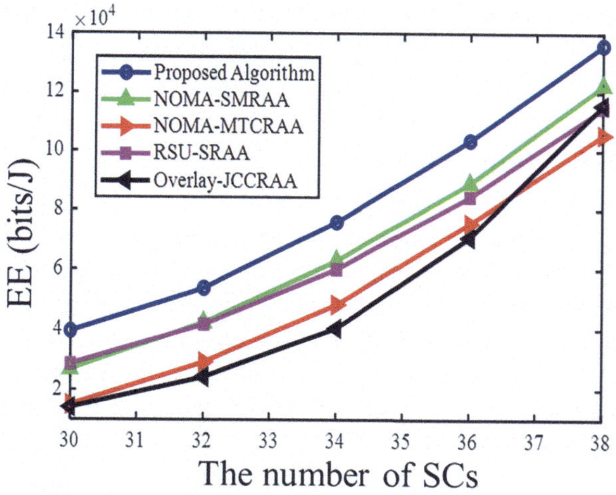

Fig. 5.4 System EE performance for different numbers of SCs

that the proposed algorithm performs best. Figure 5.3 shows the effect of the size of the offloaded tasks at each T-VU on the EE performance when the size of the offloaded tasks at each T-VU varies. The simulation results reveal that as the size of the offloaded tasks increases, the EE decreases. This is attributed to an increase in task delay, making it difficult to satisfy the delay constraint and ultimately reducing the EE performance of the underlying C-V2X network. Figure 5.4 shows that the EE increases when the number of available SCs increases for all the simulated algorithms. This is because when the number of available SCs increases, more users can occupy the spectrum bands individually, improving the system EE. Figure 5.5 shows that the curve of the social-mobility-aware algorithm (i.e., the proposed algorithm, NOMA-SMRAA, and Overlay-JCCRAA) remains stable as the number of untrusted S-VUs increases, whereas the curve of the NOMA-MTCRAA decreases

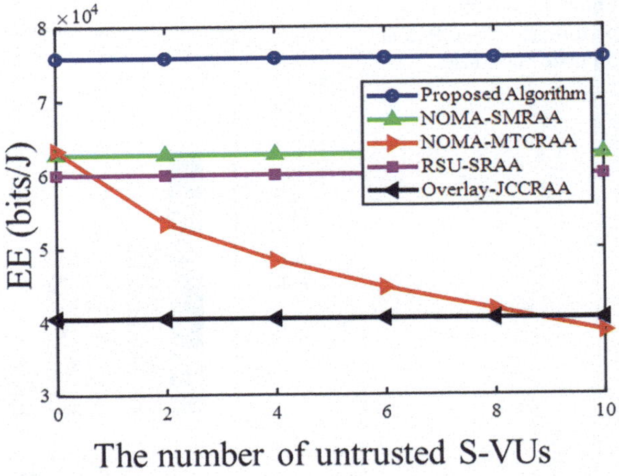

Fig. 5.5 The system EE performance for different numbers of untrusted S-VUs

significantly. This confirms that the proposed social-mobility-aware scheme is more effective than the mobility-aware scheme.

5.5 Summary

This chapter focuses on social-mobility-aware EE maximization in underlying C-V2X networks, where T-VUs can offload computational tasks to S-VUs and RSUs via NOMA. An EE maximization problem was formulated to jointly assign the 2C resources. Since the optimization was an MINLP, to solve it, an iterative JCCRAA was proposed. The simulation results show that the proposed JCCRAA not only prevents task offloading to untrusted users but also improves resource utilization and system EE.

References

1. Sun, Z., Sun, G., Liu, Y., Wang, J., & Cao, D. (2024). BARGAIN-MATCH: A game theoretical approach for resource allocation and task offloading in vehicular edge computing networks. *IEEE Transactions on Mobile Computing, 23*(2), 1655–1673.
2. Cheng, Y., Liang, C., Chen, Q., & Yu, F. R. (2021). Energy-efficient D2D-assisted computation offloading in NOMA-enabled cognitive networks. *IEEE Transactions on Vehicular Technology, 70*(12), 13441–13446.
3. Ron, D., & Lee, J.-R. (2022). DNN-based dynamic transmit power control for V2V communication underlaid cellular uplink. *IEEE Transactions on Vehicular Technology, 71*(11), 12413–12418.
4. Ron, D., Han, E., & Lee, J.-R. (2023). Spectral efficiency maximization for V2V communication underlaid cellular uplink using deep neural networks. In *2023 International Conference on Information Networking (ICOIN), Bangkok, Thailand* (pp. 569–573).

5. Luo, X., Wen, J., Kang, J., Nie, J., Xiong, Z., Zhang, Y., Yang, Z., & Xie, S. (2023). Privacy attacks and defenses for digital twin migrations in vehicular metaverses. *IEEE Network*, 1.
6. Alnasser, A., Sun, H., & Jiang, J. (2020). Recommendation-based trust model for vehicle-to-everything (v2x). *IEEE Internet of Things Journal, 7*(1), 440–450.
7. Xu, M., Niyato, D., Kang, J., Xiong, Z., & Chen, M. (2023, May). Learning-based sustainable multi-user computation offloading for mobile edge quantum computing. In *ICC 2023 – IEEE International Conference on Communications* (pp. 4045–4050).
8. Zhao, J., Liu, Y., Chai, K. K., Chen, Y., & Elkashlan, M. (2017). Joint subchannel and power allocation for noma enhanced d2d communications. *IEEE Transactions on Communications, 65*(11), 5081–5094.
9. Feng, W., Lin, S., Zhang, N., Wang, G., Ai, B., & Cai, L. (2023). Joint c-v2x based offloading and resource allocation in multi-tier vehicular edge computing system. *IEEE Journal on Selected Areas in Communications, 41*(2), 432–445.

Chapter 6
Trusted Relay-Based Physical Layer Secure Transmission

To address the problem of energy-constrained relay systems, promising solutions employing radio-frequency (RF) signal-based energy harvesting have been proposed to provide a stable energy supply for relay nodes [1]. Since RF signals can carry both information and energy, the relay node can simultaneously help transmit information and harvest energy from the signal emitted by a transmitter to prolong its lifetime, i.e., simultaneous wireless information and power transfer (SWIPT). This scheme has recently emerged as an invaluable way to meet the energy demands of energy-constrained systems and has drawn increasing attention in secure systems. In [2], a wireless-powered destination with a full-duplex (FD) architecture was proposed to enable secure transmission in SWIPT systems. The destination can convert the energy harvested from the source into the energy of artificial noise to degrade the eavesdropper channel. Additionally, in the presence of potential eavesdropping from energy harvesting (EH) nodes, an FD-capable base station was designed in [3] to achieve secure uplink and downlink transmissions concurrently and provide an energy supply to EH nodes. Compared with half-duplex (HD) schemes for cellular networks, FD relays offer a 33–65% rate gain on average. Previous works have focused on point-to-point transmissions, and secure transmissions in wireless-powered FD relay systems remain unexplored.

The secure transmission schemes and communication approaches proposed in the above works cannot be applied in wireless-powered FD relay systems directly because of different system models. Moreover, the main challenge in wireless-powered FD relay systems is that although the energy harvested at relay nodes can improve the performance of information transmission, increased information leakage and increased loopback interference (LI) can result. Unfortunately, this results in a degradation of secure performance. In addition, in a wireless-powered FD relay system, if the AF relaying protocol is employed, the LI signal strength will be amplified before being retransmitted to the destination, which worsens the above problems.

Motivated by the aforementioned problems, this work investigates the secure transmission of wireless-powered FD relay systems. Considering that FD relay

H. Zhang et al., *Intelligent Resources Management for Vehicular Social Networks*, Wireless Networks, https://doi.org/10.1007/978-3-031-80169-3_6

schemes consume less energy than FD jamming schemes while achieving the same transmission rate at the destination, we adopt FD relay schemes in this work. In particular, we consider the separate antenna configurations at the FD relay, i.e., one transmits antenna and one receive antenna, and the LI between the two antennas is considered. When the time switching (TS) protocol is adopted, the entire transmission process is divided into two phases. That is, the FD relay first harvests energy from the energy signal emitted by a multiple-antenna source and then helps with information transmission between the source and the destination with the harvested power. The main objective of this work is to investigate the effect of FD relay transmission on the system secrecy rate and characterize the fundamental trade-off between the energy harvesting time and secure communication time.

6.1 Physical Layer Secure Transmission Model

We consider the vehicle task secure transmission in a social-awarded wireless-powered full-duplex (FD) relay system, as shown in Fig. 6.1. In the social domain, this section uses the same social model as in Sect. 5.1.1 to find the eavesdropper. In the physical domain, a source transmits the confidential signal to a destination with the help of an FD relay while keeping the information secret from the eavesdropper. The source is equipped with N antennas, and other nodes are equipped with a single antenna due to volume limitations. To enable FD operation, the relay has two antennas, one for receiving and the other for transmitting. It is also assumed that the relay has a limited power supply and relies on external charging by harvesting energy from the source.

In this work, a time-switching (TS) protocol is adopted to achieve simultaneous wireless information and power transfer (SWIPT) and provide an energy supply for the relay node; hence, the entire transmission block time T is divided into two phases, as shown in Fig. 6.2a. Specifically, the α fraction of T is devoted to energy

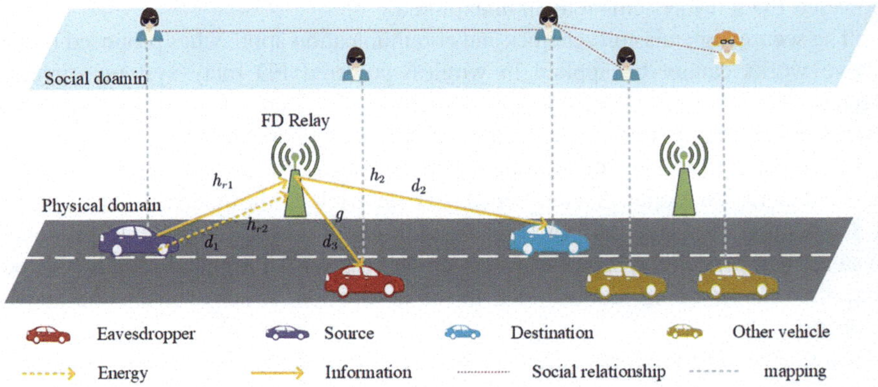

Fig. 6.1 System model of vehicle task secure transmission in wireless-powered FD relay systems

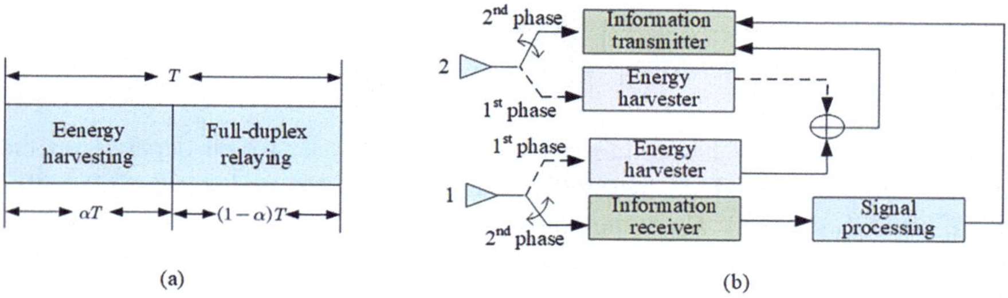

Fig. 6.2 (**a**) Two-phase SWIPT protocol. (**b**) Time switching architecture at the FD relay

harvesting, and the remaining time $(1 - \alpha)T$ is exploited for information transmission. Note that the power splitting (PS) protocol is not considered in this work since its harvested energy can be used only at the next transmission block in FD relay systems. In addition, as it is assumed that the relay node is an idle legitimate user instead of a dedicated relay node, which is selfish and depletes energy storage, it does not want to spend its own energy to support the communications of others. Therefore, TS is adopted.

In addition, to fully exploit the available hardware resource (antenna element) of the FD relay to capture more energy, both the receive antenna and the transmit antenna can be switched to an energy harvesting circuit for the first phase, as shown in Fig. 6.2b. To facilitate the following mathematical operations, we define an index term ζ to indicate whether one or two antennas are switching for energy harvesting. In particular, for $\zeta = 0$, the system works in single-antenna mode (only the receive antenna is exploited), whereas for $\zeta = 1$, the system works in multiple-antenna mode (both the receive antenna and transmit antenna are used). Therefore, the received energy signal at the relay in the first phase can be expressed as

$$\mathbf{y}_{r,H} = \frac{1}{\sqrt{d_1^m}} \begin{pmatrix} \mathbf{h}_{r1}^{\dagger} \\ \zeta \mathbf{h}_{r2}^{\dagger} \end{pmatrix} \mathbf{w}_H x_p + \mathbf{n}_{r,H}, \tag{6.1}$$

where $\mathbf{w}_H$ represents the energy beamformer and where x_p represents the energy signal with power P_s, $\mathbb{E} = \left\{ \|\mathbf{w}_H\|^2 \right\} = 1$, and $\mathbb{E} = \left\{ |x_p|^2 \right\} = P_s$, where $\mathbf{h}_{r1}$ and $\mathbf{h}_{r2}$ are the Rayleigh fading channel coefficients from the source to the receive antenna and the transmit antenna at the relay, respectively. d_1 represents the source relay distance, m is the path loss exponent, and $\mathbf{n}_{r,\ H}$ is the additive white Gaussian noise (AWGN) at the relay. When $\zeta = 0$, both $y_{r,\ H}$ and $n_{r,\ H}$ are scalars, $n_{r,H} \sim C\mathcal{N}(0, N_0)$; otherwise, when $\zeta = 1$, both $y_{r,\ H}$ and $n_{r,\ H}$ are vectors, $y_{r,\ H} \in \mathbb{C}^{2 \times 1}$, $n_{r,H} \sim C\mathcal{N}(0, N_0 \mathbf{I}_2)$.

It is assumed that the system works on the harvest–use architecture [4, 5]. This means that the energy harvested during this phase is stored in a supercapacitor and then fully consumed by the relay when the source signal is forwarded to the destination. Then, the transmit power at the relay node is denoted as

$$P_r = \frac{\eta\alpha}{1-\alpha} P_s \frac{\left|\mathbf{h}_{r1}^{\dagger}\mathbf{w}_H\right|^2 + \zeta^2\left|\mathbf{h}_{r2}^{\dagger}\mathbf{w}_H\right|^2}{d_1^m}, \tag{6.2}$$

where η is the energy harvesting efficiency, $0 < \eta < 1$, which depends on the rectification process and the energy harvesting circuitry [6], and α is the TS coefficient, where $0 < \alpha < 1$. Here, the power of the noise is not considered because its level (including both the antenna noise and the rectifier noise) is much lower than that of the energy signal [7].

In the second phase, the information signal is transmitted securely with the help of the FD relay. At the nth time block, the received signal at the FD relay is expressed as

$$y_r[n] = \frac{\mathbf{h}_{r1}^{\dagger}}{\sqrt{d_1^m}} \mathbf{w}_s x_s[n] + f x_r[n] + n_r[n]. \tag{6.3}$$

On the right-hand side of (6.3), the first part is the desired information signal, where $\mathbf{w}_s$ is the information beamformer with $\mathbb{E}\left\{\|\mathbf{w}_s\|^2\right\} = 1$ and where x_s is the information signal with $\mathbb{E}\left\{|x_s|^2\right\} = P_s$. The second part is the LI signal between the transmit antenna and receive antenna at the FD relay, where f represents the LI channel between the transmit and receive antennas of the FD relay and where $x_r[n]$ is the interference signal with $\mathbb{E}\left\{|x_r|^2\right\} = P_r$. The last part $n_r[n]$ is the zero-mean AWGN with variance N_0. Since the relay can be aware of its own signal $x_r[n]$, interference cancellation methods (i.e., analog/digital cancellation) can be applied to mitigate the LI to a tolerable level [8–10]. Thus, the post cancellation signal at the relay can be further expressed as

$$\hat{y}_r[n] = \frac{\mathbf{h}_{r1}^{\dagger}}{\sqrt{d_1^m}} \mathbf{w}_s x_s[n] + \hat{f}\hat{x}_r[n] + n_r[n], \tag{6.4}$$

where $\hat{f}\hat{x}_r[n]$ denotes the residual LI with $\mathbb{E}\left\{|\hat{x}_r[n]|^2\right\} = P_r$ and where $\hat{f}$ represents the residual LI channel caused by imperfect interference cancellation, which is normally regarded as a Rayleigh fading channel [11].

Adopting the AF protocol, the received signal in (6.4) is amplified by a factor of $\beta > 0$ ($\hat{x}_r[n] = \beta\hat{y}_r[n-1]$) and then forwarded to the destination. Note that since the processing delay is comparably small, it is ignored in the following operations. Thus, the received signals at the destination and the eavesdropper from the FD relay are expressed as

$$y_d[n] = \frac{h_2}{\sqrt{d_2^m}} x_r[n] + n_d[n] = \frac{h_2}{\sqrt{d_2^m}} \beta \left(\frac{\mathbf{h}_{r1}^{\dagger}}{\sqrt{d_1^m}} \mathbf{w}_s x_s[n] + \hat{f} \hat{x}_r[n] + n_r[n] \right) + n_d[n], \tag{6.5}$$

and

$$y_e[n] = \frac{g}{\sqrt{d_3^m}} x_r[n] + n_e[n] = \frac{g}{\sqrt{d_3^m}} \beta \left(\frac{\mathbf{h}_{r1}^{\dagger}}{\sqrt{d_1^m}} \mathbf{w}_s x_s[n] + \hat{f} \hat{x}_r[n] + n_r[n] \right) + n_e[n], \tag{6.6}$$

respectively. In (6.5), h_2 and d_2 represent the channel coefficient and the distance between the FD relay and the destination, whereas in (6.6), g and d_3 denote the channel and the distance between the FD relay and the eavesdropper, respectively. Both n_d and n_e in the above expressions are zero-mean AWGN with variance N_0. Since the average transmit power for data relaying at the relay node is P_r, the amplification factor β is expressed as

$$\beta = \sqrt{\frac{P_r}{P_s \left| \mathbf{h}_{r1}^{\dagger} \mathbf{w}_s \right|^2 / d_1^m + \left| \hat{f} \right|^2 P_r + N_0}}. \tag{6.7}$$

By substituting (6.7) into (6.5) and (6.6), the signal-to-interference-plus-noise ratio (SINR) at the destination and the eavesdropper can be obtained. They are

$$\gamma_D = \frac{\frac{P_s \left| \mathbf{h}_{r1}^{\dagger} \mathbf{w}_s \right|^2 |h_2|^2}{d_1^m d_2^m}}{\frac{P_s N_0 \left| \mathbf{h}_{r1}^{\dagger} \mathbf{w}_s \right|^2}{P_r d_1^m} + \frac{N_0^2}{P_r} + \frac{P_r \left| \hat{f} \right|^2 |h_2|^2}{d_2^m} + \frac{N_0 |h_2|^2}{d_2^m} + \left| \hat{f} \right|^2 N_0}, \tag{6.8}$$

and

$$\gamma_E = \frac{\frac{P_s \left| \mathbf{h}_{r1}^{\dagger} \mathbf{w}_s \right|^2 |g|^2}{d_1^m d_3^m}}{\frac{P_s N_0 \left| \mathbf{h}_{r1}^{\dagger} \mathbf{w}_s \right|^2}{P_r d_1^m} + \frac{N_0^2}{P_r} + \frac{P_r \left| \hat{f} \right|^2 |g|^2}{d_3^m} + \frac{N_0 |g|^2}{d_3^m} + \left| \hat{f} \right|^2 N_0}. \tag{6.9}$$

Consequently, the received information rates at the destination and the eavesdropper can be expressed as

$$R_d = \frac{\eta}{\eta + k} \log_2 \left(1 + \frac{\frac{P_s}{N_0} \frac{\left|\mathbf{h}_{r1}^{\dagger} \mathbf{w}_s\right|^2 |h_2|^2}{d_1^m d_2^m}}{\frac{\left|\mathbf{h}_{r1}^{\dagger} \mathbf{w}_s\right|^2}{k\left(\left|\mathbf{h}_{r1}^{\dagger} \mathbf{w}_H\right|^2 + \zeta^2 \left|\mathbf{h}_{r2}^{\dagger} \mathbf{w}_H\right|^2\right)} + \frac{N_0}{P_s} \frac{d_1^m}{k\left(\left|\mathbf{h}_{r1}^{\dagger} \mathbf{w}_H\right|^2 + \zeta^2 \left|\mathbf{h}_{r2}^{\dagger} \mathbf{w}_H\right|^2\right)} + Q_1 + \frac{|h_2|^2}{d_2^m} + \left|\hat{f}\right|^2} \right), \tag{6.10}$$

$$R_e = \frac{\eta}{\eta + k} \log_2 \left(1 + \frac{\frac{P_s}{N_0} \frac{\left|\mathbf{h}_{r1}^{\dagger} \mathbf{w}_s\right|^2 |g|^2}{d_1^m d_3^m}}{\frac{\left|\mathbf{h}_{r1}^{\dagger} \mathbf{w}_s\right|^2}{k\left(\left|\mathbf{h}_{r1}^{\dagger} \mathbf{w}_H\right|^2 + \zeta^2 \left|\mathbf{h}_{r2}^{\dagger} \mathbf{w}_H\right|^2\right)} + \frac{N_0}{P_s} \frac{d_1^m}{k\left(\left|\mathbf{h}_{r1}^{\dagger} \mathbf{w}_H\right|^2 + \zeta^2 \left|\mathbf{h}_{r2}^{\dagger} \mathbf{w}_H\right|^2\right)} + Q_2 + \frac{|g|^2}{d_s^m} + \left|\hat{f}\right|^2} \right), \tag{6.11}$$

where $Q_1 = k \frac{P_s}{N_0} \frac{\left(\left|\mathbf{h}_{r1}^{\dagger} \mathbf{w}_H\right|^2 + \zeta^2 \left|\mathbf{h}_{r2}^{\dagger} \mathbf{w}_H\right|^2\right) \left|\hat{f}\right|^2 |h_2|^2}{d_1^m d_2^m}$, $Q_2 = k \frac{P_s}{N_0} \frac{\left(\left|\mathbf{h}_{r1}^{\dagger} \mathbf{w}_H\right|^2 + \zeta^2 \left|\mathbf{h}_{r2}^{\dagger} \mathbf{w}_H\right|^2\right) \left|\hat{f}\right|^2 |g|^2}{d_1^m d_3^m}$. κ denotes the transmission time ratio of energy harvesting and information transmission and is defined as $k \triangleq \eta\alpha/(1 - \alpha)$.

The achievable secrecy rate is defined as [12]

$$R_s = [R_d - R_e]^+. \tag{6.12}$$

The detailed analytical expression of the secrecy rate is obtained by substituting (6.10) and (6.11) into (6.12).

6.2 Secrecy Rate Maximization Problem Formulation

The objective of this work is to jointly optimize the beamformers and the TS coefficient to maximize the secrecy rate performance, subject to the given power constraints at the source for both energy harvesting and information transmission. The optimization problem is hence formulated as

$$\begin{aligned} \text{P1}: \max_{\mathbf{w}_H, \mathbf{w}_s, \alpha} \; & R_s(\mathbf{w}_H, \mathbf{w}_s, \alpha) \\ \text{s.t.} & \begin{cases} \|\mathbf{w}_H\|^2 \leq 1 \\ \|\mathbf{w}_s\|^2 \leq 1 \\ 0 < \alpha < 1 \end{cases} \end{aligned} \tag{6.13}$$

Since the beamforming vectors $\mathbf{w}_H$ and $\mathbf{w}_s$ and the TS coefficient α are coupled, the formulated optimization problem is easily found to be nonconvex. Therefore, the complexity of finding the optimal solution is prohibitively high, and it is almost impossible to concurrently solve the three variables ($\mathbf{w}_H$, $\mathbf{w}_s$, α). As an alternative, a suboptimal algorithm, i.e., an iterative algorithm, is proposed to solve the formulated problem P1.

6.3 Iterative Algorithm for Secrecy Rate Optimization

As mentioned, instead of seeking the optimal solution for P1, we attempt to develop a suboptimal scheme to solve the optimization problem efficiently, i.e., optimizing the energy beamforming vector together with the information beamforming vector and the TS coefficient iteratively. This idea is implemented because the secrecy rate expression (6.12) is a concave function of $\mathbf{w}_H$ while keeping $\mathbf{w}_s$ and α fixed, and vice versa. If the iteration method converges, a suboptimal solution can be obtained for the original optimization problem P1. In this part, the beamforming vectors and TS coefficient design, together with the convergence property, are successively analyzed.

6.3.1 Beamforming Design

With the fixed information beamforming vector $\mathbf{w}_s$ and the fixed TS coefficient α, the secrecy rate is maximized by designing a proper energy beamforming vector, which is expressed as

$$\begin{aligned} \text{P2.1}: \quad & \max_{\mathbf{w}_H} R_s(\mathbf{w}_H) \\ \text{s.t.} \quad & \|\mathbf{w}_\text{H}\|^2 \leq 1. \end{aligned} \tag{6.14}$$

For the optimal solution of P2.1, we have the following proposition.

Proposition 1 The optimal solution to problem (6.14) is achieved as

$$\mathbf{w}_H^* = \begin{cases} \sqrt{\mu/\lambda_\text{max}}\,\mathbf{q}^\text{unit}, & 0 < \mu/\lambda_\text{max} \leq 1 \\ \mathbf{q}^\text{unit}, & \text{otherwise}, \end{cases} \tag{6.15}$$

where $\mathbf{q}^\text{unit}$ is the unit-norm eigenvector corresponding to the maximum eigenvalue of matrix $[\mathbf{h}_{r1}, \zeta\mathbf{h}_{r2}]\ [\mathbf{h}_{r1}, \zeta\mathbf{h}_{r2}]^\dagger$ and where μ is a scalar chosen to adjust the power of the energy beamforming vector in (6.15) to satisfy the power constraint in (6.14). In terms of whether the LI is eliminated completely, μ has two different expressions. When the LI is removed completely, the mean of $|\hat{f}|^2$ equals zero, $\lambda_f = 0$, and the parameter μ is written as

$$\mu = \frac{N_0 d_1^m}{P_s k} \sqrt{\frac{P_s \left| \mathbf{h}_{r1}^{\dagger} \mathbf{w}_s \right|^2 / d_1^m}{N_0 |h_2|^2 |g|^2 / (d_2^m d_3^m)}}. \tag{6.16}$$

Otherwise, when the residual LI is not fully eliminated, the parameter μ can be evaluated numerically.

Then, with the above-obtained energy beamformer $\mathbf{w}_H$ and the given fixed TS coefficient α, the information beamforming vector $\mathbf{w}_s$ can be designed by maximizing the secrecy rate

$$\begin{aligned} \text{P2.2}: \quad & \max_{\mathbf{w}_s} R_s(\mathbf{w}_s) \\ \text{s.t.} \quad & \|\mathbf{w}_s\|^2 \le 1. \end{aligned} \tag{6.17}$$

On the basis of the analytical expression of R_s, the information beamforming vector $\mathbf{w}_s$ is always coupled with the source-to-relay channel coefficient $\mathbf{h}_{r1}$; thus, the secrecy rate performance of $\left|\mathbf{h}_{r1}^{\dagger} \mathbf{w}_s\right|^2$ is first investigated. Under the positive secrecy rate constraint $|h_2|^2/d_2^m \ge |g|^2/d_3^m$, verified in Proposition 4 of Appendix B, the secrecy rate $R_s\left(\left|\mathbf{h}_{r1}^{\dagger} \mathbf{w}_s\right|^2\right)$ is an increasing function of $\left|\mathbf{h}_{r1}^{\dagger} \mathbf{w}_s\right|^2$. Therefore, the original problem P2.2 is equal to maximizing $\left|\mathbf{h}_{r1}^{\dagger} \mathbf{w}_s\right|^2$, with the power constraint $\|\mathbf{w}_s\|^2 \le 1$, i.e.,

$$\begin{aligned} P2.2.1: \quad & \max_{\mathbf{w}_s} \mathbf{w}_s^{\dagger} \mathbf{h}_{r1} \mathbf{h}_{r1}^{\dagger} \mathbf{w}_s \\ s.t. \quad & \|\mathbf{w}_s\|^2 \le 1. \end{aligned} \tag{6.18}$$

The optimal solution is derived as

$$\mathbf{w}_s^* = \frac{\mathbf{h}_{r1}}{\left\|\mathbf{h}_{r1}\right\|}. \tag{6.19}$$

Regardless of whether the LI is eliminated completely, the information transmission can be seen as an MISO system, so it is obvious that the solution is identical to that of a conventional MISO transmission with maximum ratio transmission (MRT).

Algorithm 6.1: Iterative Algorithm

Initialize $\mathbf{w}_H^0$, $\mathbf{w}_s^0$, α^0, ε, calculate $R_s\left(\mathbf{w}_H^0, \mathbf{w}_s^0, \alpha^0\right)$ and set $n = 0$.

Repeat

$n = n + 1$

For the fixed $\mathbf{w}_s^{n-1}$, α^{n-1}, find the solution of $\mathbf{w}_H^n$ using Proposition 1.

With $\mathbf{w}_H^n$ and fixed α^{n-1}, calculate the solution of $\mathbf{w}_s^n$ using (6.19).

For the obtained $\mathbf{w}_H^n$ and $\mathbf{w}_s^n$, find the numerical solution of α^n, and calculate the secrecy rate $R_s^n = R_s(\mathbf{w}_H^n, \mathbf{w}_s^n, \alpha^n)$.

Until $|R_s^n - R_s^{n-1}| \leqslant \varepsilon$.

Return $(\mathbf{w}_H^*, \mathbf{w}_s^*, \alpha^*) = (\mathbf{w}_H^n, \mathbf{w}_s^n, \alpha^n)$

6.3.2 Time Switching Coefficient Design

With the obtained $\mathbf{w}_H$ and $\mathbf{w}_s$ in the above two parts, the TS coefficient α can be obtained by maximizing the secrecy rate

$$\begin{aligned} \text{P2.3}:\ &\max_{\alpha} R_s(\alpha) \\ \text{s.t.}\ \ &0 < \alpha < 1. \end{aligned} \tag{6.20}$$

Proposition 2 With known $\mathbf{w}_H$ and $\mathbf{w}_s$, problem (6.20) is concave on α.

Since problem (6.20) is concave, a unique optimal solution to P2.3 always exists in the interval $\alpha \in (0,\ 1)$ and can be obtained easily via numerical methods, including the Newton method [13] and bisection method [14].

By alternatively updating the beamforming vectors obtained via (6.15) and (6.19) as well as the TS coefficient obtained via (6.20), the iterative algorithm is summarized in Algorithm 6.1.

6.4 Simulations and Discussion

To validate the analytical expressions developed in the previous sections and investigate the secrecy rate performances for the wireless-powered FD relay scheme, Monte Carlo simulations are performed in this section. All the simulation results are obtained by averaging over 10^4 independent trials. Unless otherwise specified, the source-relay distance d_1 and the relay-destination distance d_2 are both normalized to unit values [10], and the relay-eavesdropper distance d_3 is set equal to $1.2d_2$. In addition, the mean values of $\|\mathbf{h}_{r1}\|^2$ and $\|\mathbf{h}_{r2}\|^2$ are both set to 1, i.e., $\lambda_d = 1$. Similarly, the mean values of $|g|^2$ and $|\hat{f}|^2$ are set to $\lambda_e = 1$ and $\lambda_f = 0.2$, respectively. The path loss exponent is also assumed to be $m = 3$, the energy harvesting efficiency is $\eta = 0.6$, and the number of antennas at the source is $N = 5$.

6.4.1 Concavity Analysis

As analyzed in the previous sections, the secrecy rate is a concave function of α. We simulate this concavity and the corresponding secrecy rate in this part. In doing so, we focus on a single transmission block with the following randomly generated channel coefficients:

$$[\mathbf{h}_{r1}, \mathbf{h}_{r2}] = \begin{pmatrix} 0.7409 - i0.3587 & -0.7906 + i0.4592 \\ 1.4558 - i1.0487 & -0.4221 + i0.4232 \\ 0.0189 - i0.0753 & -0.4805 - i0.1361 \\ -0.71701 + i0.1734 & 0.6753 - i1.4006 \\ 0.7343 + i0.1031 & -0.0925 + i0.2566 \end{pmatrix}.$$

In addition, it is assumed that $h_2 = 0.4978 - i0.5322$, $g = -0.0120 - i0.3123$, and $f = -0.1709 + i0.3167$, which are also randomly generated following their statistical properties. Without loss of generality, we set $P_s/N_0 = 10$ dB. The results are included in Fig. 6.3 and show that the simulations are highly concurrent with the theoretical analysis. Therefore, there exists one and only one optimal TS coefficient α^* maximizing R_s, which has also been proven in Proposition 2. In addition, since the LI at the FD relay can degrade the received rate at the destination, the secrecy rate performance of the LI-free (LI is eliminated completely) case is better than that of the

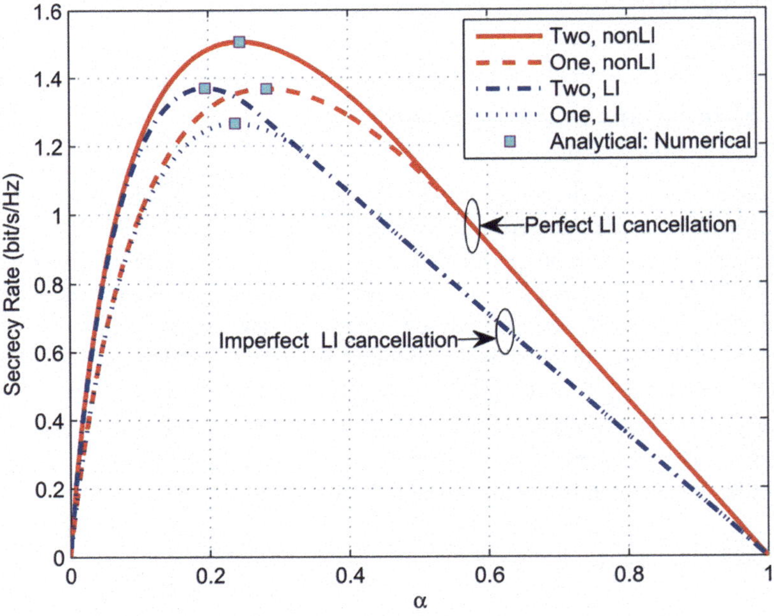

Fig. 6.3 Secrecy rate versus α. The strength of the LI is $\lambda_f = 0.2$, $P_s/N_0 = 10$ dB

imperfect interference cancellation case. From the results shown in Fig. 6.3, it can be observed that the secrecy rate performance of employing two antennas for energy harvesting outperforms that of employing only one antenna in the low-α regime, whereas in the high-α regime, the performances of the two are quite similar. This is because when α is small, two antennas can harvest more energy to facilitate information transmission than one antenna can. When α is large, an excessive amount of energy is collected, which is actually detrimental since this causes more information leakage at the eavesdropper and strong LI. This will ultimately reduce the secrecy rate.

6.4.2 Convergence Performance

An iterative algorithm is proposed to solve the formulated optimization problem in Sect. 6.3. The effectiveness of the proposed algorithm depends on its convergence. If the iterative algorithm converges, then a suboptimal solution can be obtained. To investigate the convergence property, we choose the previously described two-antenna case as an example and simulate the system performance. In the simulations shown in Fig. 6.4a, the same channel coefficients used to obtain Fig. 3 are employed to achieve the convergence performance of the secrecy rate in one transmission block. Figure 6.4a shows that after 3 iterations, the secrecy rate converges and approaches the analytical upper bound of the achievable secrecy rate with a gap of 0.0002 bits/s/Hz.

To further demonstrate the fast convergence property of the proposed iterative algorithm, the convergence performances over 1000 different transmission time blocks are simulated and listed in Fig. 6.4b. The channel coefficients over blocks are statistically independent and identical. The overall average number of iterations is 3.84. Thus, both Fig. 6.4a and b demonstrate that *Proposition 3* is valid from the simulation perspective and that the proposed iterative algorithm has fast convergence speed.

6.4.3 Key System Parameters

In wireless-powered secure communication systems, increasing P_s/N_0 has both constructive and destructive effects on system performance. An increase in P_s/N_0 helps the relay harvest more energy and consequently improves the received SINR at the destination. Moreover, increasing P_s/N_0 increases the information signal strength from the source, which leads to serious information leakage at the eavesdropper and LI at the relay. The secrecy rate performance for different P_s/N_0 values is shown in Fig. 6.5 with an optimized α^* and fixed $\alpha = \{0.01,\ 0.5,\ 0.9\}$. Without loss of generality, the strength of the LI is set to $\lambda_f = 0.2$. It is obvious that increasing P_s/N_0 can improve the secrecy rate. In the lower P_s/N_0 regime, since employing two

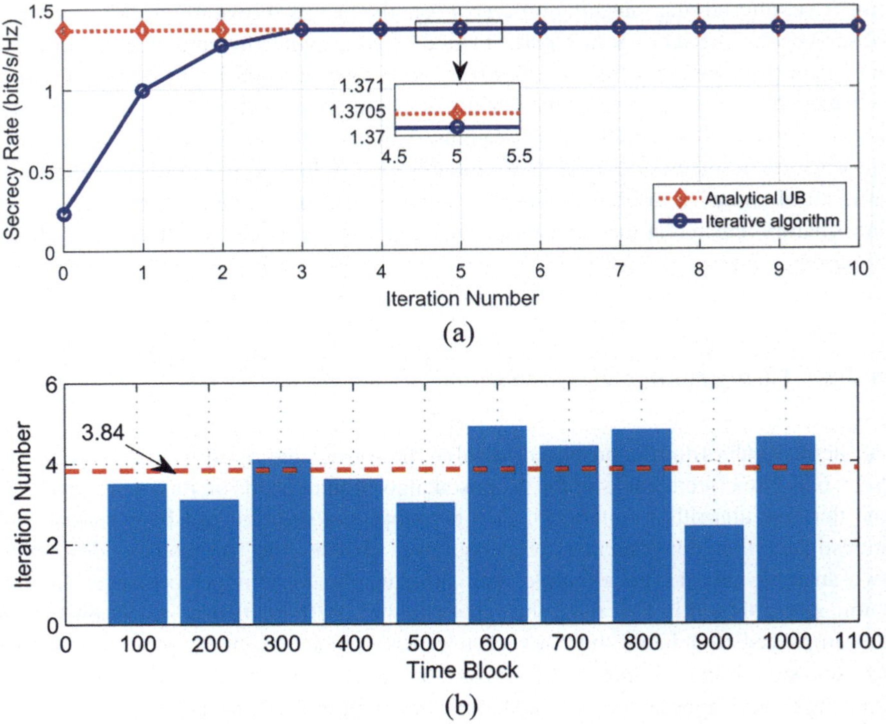

Fig. 6.4 Convergence performance of the iterative algorithm (Algorithm 1). Two antennas are used for energy harvesting at the FD relay, and the strength of the LI is $\lambda_f = 0.2$, $P_s/N_0 = 10$ dB. (**a**) Secrecy rate versus the number of iterations. (**b**) The average number of iterations during 1000-time blocks

antennas at the FD relay yields more energy than employing one antenna, it achieves a better secrecy rate performance. As P_s/N_0 increases, more energy is harvested at the relay, resulting in more information leakage at the eavesdropper and serious LI at the FD relay. As a result, the increase in the achievable secrecy rate will be slow. In addition, because a fixed α is not the optimal solution, it may lead to transmission rate loss or more information leakage and severe interference. Thus, the secure performance for a fixed α is worse than that with the optimized α^*. Specifically, in the high-P_s/N_0 regime, adopting an optimized α^* results in more than a 96.7% secrecy rate gain compared with that of the fixed $\alpha = 0.5$ and $\alpha = 0.9$ cases. In the low-P_s/N_0 regime, an optimized α^* also results in more than a 73.5% increase in the secrecy rate compared with that when $\alpha = 0.01$ and $\alpha = 0.9$ are fixed. This finding demonstrates that the optimized α^* can substantially increase the secrecy rate and that optimizing the time switching coefficient α is necessary.

The energy conversion efficiency factor η determines how much of the received power the relay can actually harvest. A higher η allows the relay to harvest more energy, which in turn increases the transmit power at the relay and results in an

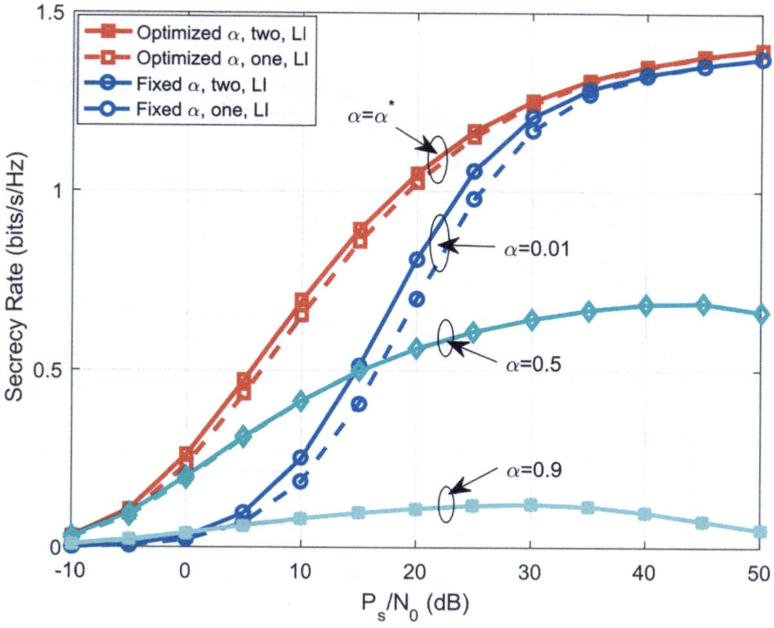

Fig. 6.5 Secrecy rate versus P_s/N_0, with optimized α^* and fixed α

enhanced received SINR at the destination, ultimately increasing the secrecy rate, as shown in Fig. 6.6. Nevertheless, as the harvested energy increases along with η, it also leads to more serious information leakage, eventually resulting in the secure performance increasing slowly in the high η regime. Moreover, compared with the secure performance of the HD relay system, even the LI is introduced, the FD relay can also significantly improve the secrecy rate. In addition, the secrecy rate for the FD relay under the high-SNR approximation is also given, and the results are concurrent with those under the exact SNR. This indicates that the approximate expression has similar monotonicity with the exact analytical expression in (6.12). This property is used in *Proposition 3* to analyze the convergence of Algorithm 1. More importantly, the results also reveal that the secrecy rate gap between the proposed method and the analytical upper bound is less than 0.02 bits/s/Hz. This finding demonstrates that the proposed iterative algorithm can achieve a near-optimal secrecy rate.

As illustrated in the above figures, increases in Ps/N0 and η can both increase the harvested energy and ultimately lead to an increase in the secrecy rate. However, the increase in the harvested energy also leads to severe LI at the FD relay, which in turn makes the increase in the secrecy rate slow in the high Ps/N0 (or η) regime. Therefore, the strength of LI is a key parameter determining the superiority of the secrecy rate performance in wireless-powered FD relay systems. In Fig. 6.7, the secrecy rate performances for different LI strengths in both FD relay and HD relay systems are given. The increasing strength of LI causes an increase in interference at

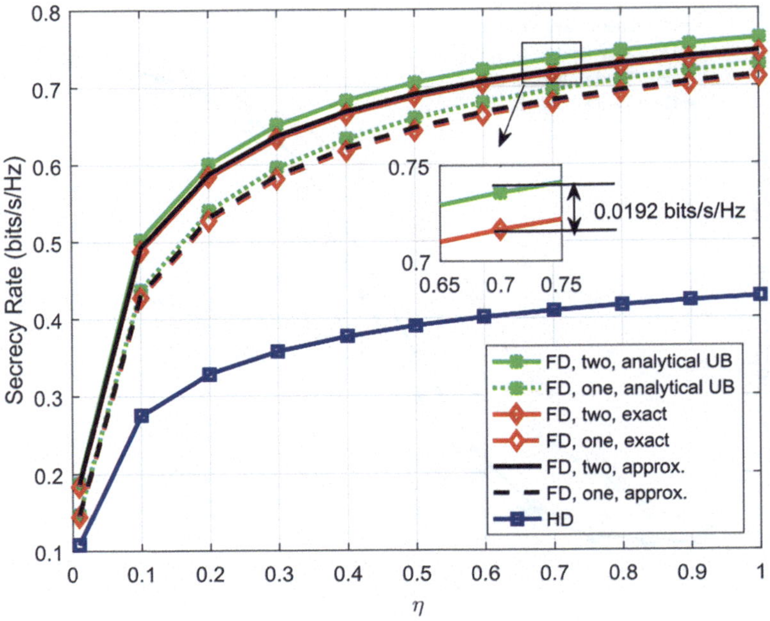

Fig. 6.6 Secrecy rate versus energy harvesting efficiency η. The secure performance of the FD relay system is compared with that of the HD relay system

the FD relay, then degrades the received SINR at the destination, finally leading to a loss in secrecy rate. However, because there is no interference in HD relay systems, the secrecy rate remains constant. However, the secure performance of FD relay systems still outperforms that of HD relay systems for any value of λ_f. In addition, compared with single-input/single-output (SISO) systems, multiple antennas equipped at the source node with energy beamforming and information beamforming can significantly improve the secrecy rate. This is intuitive since being equipped with more antennas at the source results in sharper energy and information beams, which in turn yields higher secure performance.

The relay location can also impact the received information rate at receivers (both destinations and eavesdroppers). Thus, the relay layout is highly important in cooperative secure systems. This work analyzes the secrecy rate performance versus the source–relay distance for different eavesdropper locations, and the results are included in Fig. 6.8. In contrast to a relay with conventional power supplies (such as a power grid or battery), where the optimal relay position is somewhere between the source node and the destination node to maximize the secrecy rate, a wireless-powered relay near the source or the destination can offer optimal secure performance. This difference lies in the fact that the EH relay is solely powered by the source. When the relay is near the source, more energy is harvested for secure information transmission, which helps to achieve better secure performance for both $d_2 < d_3$ and $d_2 > d_3$. As the relay moves far away from the source and

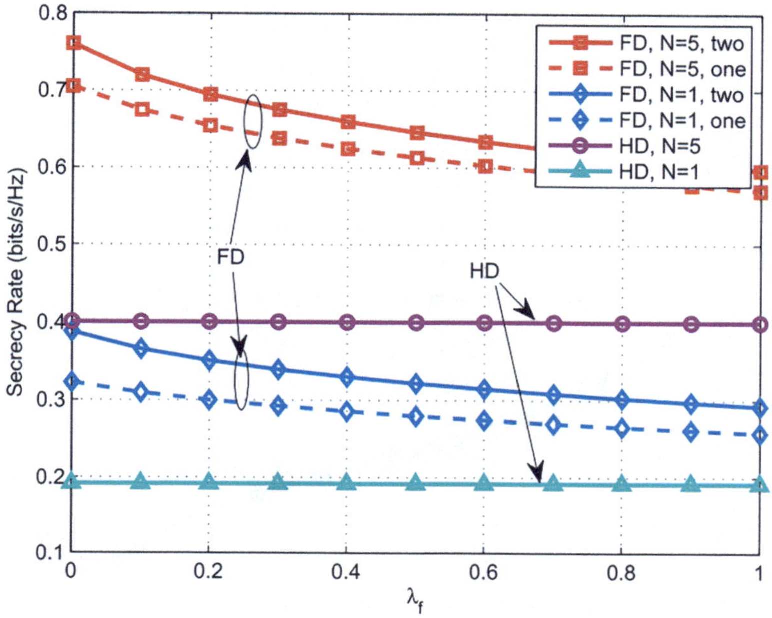

Fig. 6.7 Secrecy rate versus the strength of LI, with the optimized α obtained from the iterative algorithm

approaches the destination, although the energy harvested from the source decreases, the required transmit power for the relay to achieve information transmission from the relay to the destination is also reduced. Thus, the destination can still achieve better performance. Moreover, when the relay–eavesdropper distance is smaller than the relay–destination distance, i.e., $d_2 > d_3$, the eavesdropper can intercept much more information, leading to a significant loss in secure performance.

6.5 Summary

This chapter investigated the secrecy rate performance of wireless-powered FD relay systems. By adopting the TS protocol, the FD relay harvests energy from the source RF signals and stores it as the power supply for confidential information transmission. The objective of this work is to maximize the system secrecy rate. In doing so, an optimization problem is formulated by jointly considering the energy beamformer, the information beamformer, and the TS coefficient. As the formulated problem is nonconvex, an iterative algorithm has been proposed to solve it. The closed-form solutions of the energy beamformer and information beamformer were derived theoretically. Moreover, on the basis of the concavity of the secrecy rate on the TS coefficient, a numerical solution was obtained. Simulations were also

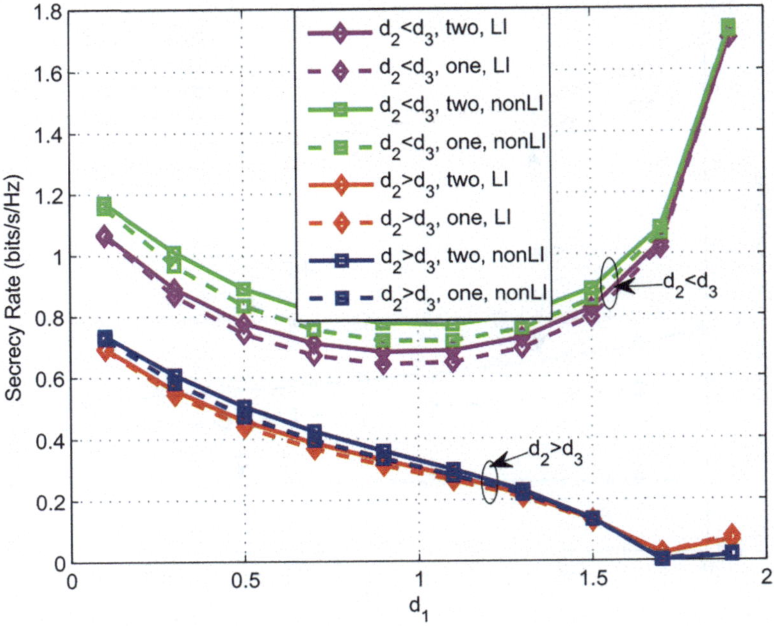

Fig. 6.8 Secrecy rate versus source–relay distance. The distance between the source and destination is assumed to be 2.0, and the distance between the source and eavesdropper is assumed to be 2.2 or 1.8

performed to verify the proposed scheme, and the results revealed that the choice of the TS coefficient plays a critical role in determining the wireless-powered FD relay gains. In addition, exploiting two antennas in the FD relay for energy harvesting can improve the harvested energy, thereby enhancing the secrecy rate performance. The results also demonstrate that the FD relaying architecture can substantially increase the secrecy rate compared with the HD relaying architecture for both one-antenna and two-antenna cases.

References

1. Li, R., Xue, J., Sun, J., & Chatzinotas, S. (2024). A deep learning approach for universal NPRACH detection with inter-cell interference. *IEEE Transactions on Communications, 72*(3), 1401–1413.
2. Li, X., Wang, W., Zhang, M., Zhou, F., & Al-Dhahir, N. (2020). Robust secure beamforming for SWIPT-aided relay systems with full-duplex receiver and imperfect CSI. *IEEE Transactions on Vehicular Technology, 69*(2), 1867–1878.
3. Ge, Y., & Fan, J. (2022). Robust secure beamforming for intelligent reflecting surface assisted full-duplex MISO systems. *IEEE Transactions on Information Forensics and Security, 17*, 253–264.

4. Zhou, Y., Jin, Z., Shi, H., Shi, L., & Lu, N. (2024). Flying IRS: QoE-driven trajectory optimization and resource allocation based on adaptive deployment for WPCNs in 6G IoT. *IEEE Internet of Things Journal, 11*(5), 9031–9046.
5. Xin, G., Hu, F., Ling, Z., Na, S., & Jin, C. (2023). Dynamic scheduling for minimizing age penalty in resource-constrained classified WBANs with energy harvesting. *IEEE Sensors Journal, 23*(19), 23638–23652.
6. Tang, J., Peng, Z., So, D. K. C., Zhang, X., Wong, K.-K., & Chambers, J. A. (2023). Energy efficiency optimization for a multiuser IRS-aided MISO system with SWIPT. *IEEE Transactions on Communications, 71*(10), 5950–5962.
7. Mou, W., Cai, Y., Yang, W., Yang, W., Xu, X., & Hu, J. (2015). Exploiting full Duplex techniques for secure communication in SWIPT system. In *2015 International Conference on Wireless Communications & Signal Processing (WCSP), Nanjing, China* (pp. 1–6).
8. Liu, Y., et al. (2021). Reconfigurable intelligent surfaces: Principles and opportunities. *IEEE Communications Surveys & Tutorials, 23*(3), 1546–1577.
9. Zhou, J., et al. (2017). Integrated full duplex radios. *IEEE Communications Magazine, 55*(4), 142–151.
10. Liu, B., Jiang, J., Liu, Q., Liu, J., & Xu, S. (2024). An intelligent reflecting surface-based attack scheme against dual-functional radar and communication systems. *IEEE Transactions on Information Forensics and Security, 19*, 4945–4956.
11. Fang, S., Chen, G., Xiao, P., Wong, K.-K., & Tafazolli, R. (2024). Intelligent omni surface-assisted self-interference cancellation for full-duplex MISO system. *IEEE Transactions on Wireless Communications, 23*(3), 2268–2281.
12. Zhang, Y., Zhang, H., Xiao, S., Tang, W., & Eldar, Y. C. (2024). Near-field wideband secure communications: An analog beamfocusing approach. *IEEE Transactions on Signal Processing, 72*, 2173–2187.
13. Wang, S., He, Z., & Rong, Y. (2021). Joint transceiver optimization for DF multicasting MIMO relay systems with wireless information and power transfer. *IEEE Transactions on Communications, 69*(7), 4953–4967.
14. Liu, Y., et al. (2022). Evolution of NOMA toward Next Generation Multiple Access (NGMA) for 6G. *IEEE Journal on Selected Areas in Communications, 40*(4), 1037–1071.

Chapter 7
Conclusions and Future Research Directions

7.1 Conclusions

In this study, we investigated intelligent resource management issues in VSNs. Specifically, we proposed a novel framework for wireless traffic prediction in VSNs, where the movement behavior of vehicles, as well as the communication behavior of vehicles, were considered jointly. On the basis of the prediction results, we proposed a social-mobility-aware radio communication resource allocation algorithm to improve the sum-rate performance in a VSN. A socially aware proactive edge caching strategy in VSNs achieved a trade-off between the cost of providing caching services and the content transmission latency. Next, we designed a joint communication and computation resource management strategy to improve the offloading EE and proposed a trusted relay-based scheme. The main contributions of this study are summarized as follows.

- A novel framework for wireless traffic prediction was proposed for VSNs, where the movement behavior of vehicles, as well as the communication behavior of vehicles, were considered jointly. We first proposed a novel method to match the movement and communication behavior of vehicles by merging two independent datasets on the trajectories of vehicles and communication traffic volumes. Then, a novel STeP-UNet was proposed, in which the STeP is embedded to capture cross-domain features of the wireless traffic pattern, and the UNet structure is utilized to realize the skipping connection from the front layer to the back layer to fuse different resolutions.
- A social-mobility-aware underlying C-V2X network was investigated. To meet the various QoS requirements of both cellular users and vehicular users and maximize the system sum rate, we proposed a jointly optimized vehicle pairing and radio resource (spectrum and power) assignment. Since the formulated optimization problem turned out to be NP-hard, an iterative method was proposed to solve it by decoupling the original problem into two subproblems, i.e., the joint vehicle pairing and spectrum assignment subproblem and the power control

H. Zhang et al., *Intelligent Resources Management for Vehicular Social Networks*, Wireless Networks, https://doi.org/10.1007/978-3-031-80169-3_7

subproblem. The HSMA-VPA and a revised KM-SAA were designed to solve the first subproblem. Given the vehicle pairing and spectrum assignment results, the second subproblem was solved via a low-complexity 3D PCA. Through iterations, the vehicle pairing variables, spectrum assignment variables, and transmit power control were iteratively updated until reaching convergence.

- A novel community detection and attention-weighted federated learning-based proactive edge caching strategy was proposed. In this strategy, we first grouped vehicles into different communities on the basis of both the mobility and social properties of the vehicles and then selected IVs as caching nodes for each community by considering the social importance of the vehicles. To determine how to place the popular content in these selected IVs, an AWFL-based content popularity prediction framework was proposed. It integrated the AWFL and BiLSTM to achieve higher content popularity prediction accuracy while protecting user privacy. Considering the imbalance of the active levels and local computing capacities of vehicles, an attention-weighted aggregation mechanism was proposed to improve the training efficiency and prediction accuracy.
- We focused on social-mobility-aware EE maximization in underlying C-V2X networks, where the T-VUs can offload the computation tasks to the S-VUs and the RSUs. An EE maximization problem was formulated to jointly assign the 2C resources. Since the optimization process turned out to be MINLP, we transformed it into separate computation and communication resource allocation subproblems to solve it. To address the first subproblem, we fully integrated the social and mobility characteristics and designed a heuristic algorithm to achieve edge server selection and task splitting. To address the complex co-channel interference in the second subproblem, the power allocation and spectrum assignment solutions were obtained via a tightening lower bound method and a KM algorithm. The simulation results showed that the proposed JCCRAA not only prevents task offloading to untrusted users but also improves resource utilization and system EE.
- We investigated the secrecy rate performance of wireless-powered FD relay systems. By adopting the TS protocol, the FD relay harvests energy from the source RF signals and stores it as the power supply for confidential information transmission. The objective of this work was to maximize the system secrecy rate. In doing so, an optimization problem was formulated by jointly considering the energy beamformer, the information beamformer, and the TS coefficient. As the formulated problem was nonconvex, an iterative algorithm was proposed to solve it. The closed-form solutions of the energy beamformer and information beamformer were derived theoretically. Moreover, on the basis of the concavity of the secrecy rate on the TS coefficient, a numerical solution was obtained. Simulations were also performed to verify the proposed scheme, and the results revealed that the choice of the TS coefficient plays a critical role in determining the wireless-powered FD relay gains. In addition, exploiting two antennas of the FD relay for energy harvesting can improve the harvested energy, thereby enhancing the secrecy rate performance.

7.2 Future Research Directions

Looking ahead, building upon current trends and emerging challenges, some future research directions for intelligent resource management in vehicular social networks can be identified:

1. *Context-aware Resource Allocation*: Develop algorithms that allocate resources dynamically on the basis of contextual information such as traffic conditions, road topology, vehicle types, and user preferences. This involves integrating data from vehicular sensors, GPSs, and external sources to optimize resource utilization.
2. *Autonomous Resource Management*: Investigate autonomous resource management techniques where vehicles dynamically negotiate and manage resources (such as bandwidth, computational power, and storage) without centralized control. This could involve game-theoretic approaches, distributed algorithms, and decentralized decision-making mechanisms.
3. *Edge Computing and Edge Intelligence*: Leverage edge computing and mobile edge intelligence to perform real-time data processing and resource allocation closer to the vehicles. Develop edge-based algorithms and frameworks for optimizing latency-sensitive applications and reducing dependency on centralized cloud resources.
4. *AI for Resource Optimization*: Advance AI and machine learning techniques to predict resource demands and optimize resource allocation in vehicular networks. Explore reinforcement learning for autonomous decision-making in dynamic and uncertain environments, considering factors such as traffic patterns and user preferences.
5. *Federated Learning for Resource Optimization*: Utilize federated learning techniques to train machine learning models collaboratively across vehicles without needing to share sensitive data. Apply federated learning to predict resource demands, optimize resource allocation, and improve decision-making in dynamic vehicular environments.
6. *Privacy-preserving Resource Management*: Develop robust privacy-preserving mechanisms for managing sensitive data and resources in vehicular social networks. This includes encryption techniques, differential privacy, and secure multiparty computation to protect user privacy while enabling efficient resource allocation.
7. *Interoperability and Standardization*: Promote interoperability among different VSNs, vehicles, and infrastructure by defining common standards and protocols for resource management. Foster collaboration between industry stakeholders to ensure the seamless integration of diverse communication technologies and facilitate efficient resource sharing.
8. *Simulation and Testbed Development*: Build comprehensive simulation models and test them to evaluate the performance of the proposed resource management strategies under various scenarios and conditions. Validate research findings

through realistic simulations and experimental deployments to demonstrate scalability, robustness, and effectiveness.

By focusing on these future research directions, researchers can advance the state of the art in intelligent resource management for vehicular social networks, ultimately contributing to safer, more efficient, and more connected smart transportation systems in the future.

Index

H. Zhang et al., *Intelligent Resources Management for Vehicular Social Networks*, Wireless Networks, https://doi.org/10.1007/978-3-031-80169-3

The manufacturer's authorised representative in the EU is Springer Nature Customer Service Centre GmbH, Europaplatz 3, 69115 Heidelberg, Germany. If you have any concerns regarding our products, please contact ProductSafety@springernature.com

Printed and bound by CPI Group (UK) Ltd, Croydon, CR0 4YY
30/09/2026
02225577-0004